THE BRAIN AND THE INNER WORLD

THE BRAIN AND THE INNER WORLD

AN INTRODUCTION TO THE NEUROSCIENCE OF SUBJECTIVE EXPERIENCE

MARK SOLMS AND OLIVER TURNBULL

FOREWORD

OLIVER SACKS

Other Press
New York

This book was set in ITC Bookman.

ISBN-13: 978-1-59051-017-9

10 9 8 7 6 5 4

 Printed in the United States of America on acid-free paper. For information write to Other Press, LLC, 2 Park Avenue, 24th Floor, New York, NY 10016. Or visit our Web site: www.otherpress.com.

Library of Congress Cataloging-in-Publication Data

Solms, Mark
The brain and the inner world: an introduction to the neuroscience of subjective experience / Mark Solms and Oliver Turnbull.
p. cm.
Includes bibliographical references and index.
ISBN 1-59051-035-6 (hbk.)—ISBN 1-59051-017-8 (pbk.)
1. Higher nervous activity. 2. Neuropsychology. 3. Subjectivity.
I. Turnbull, Oliver, 1964– II. Title.
QP395 .S65 2002
612.8'2—dc21 2002016900

CONTENTS

FOREWORD BY OLIVER SACKS

This is the fourth book of which Mark Solms is the principal author. *The Brain and the Inner World*, coauthored with Oliver Turnbull, is complementary to Solms's earlier books, expanding and clarifying considerations raised there, especially in his *Neuropsychology of Dreams* and *Clinical Studies in Neuro-Psychoanalysis*. The main ideas in these books have exercised Dr. Solms for at least fifteen years, for it was in early 1987 that he first wrote to me, enclosing a fascinating paper in which he spoke of his intention "to examine the relationship between psychoanalysis and neuropsychology . . . [and] to show that psychoanalysis is based on sound neurological principles"—grand ideals, which I could only admire.

In all of his work, Solms has clarified that so-often-misunderstood "moment of transition" in the 1890s when Freud appeared to abandon a neurological explanation for psychoanalysis (Solms's first book, coedited with Michael Saling, was entitled *A Moment of Transition*). The reason for this, Solms showed, was the very inadequate state of neurological (and physiological) understanding at the time, not any turning against neurological explanation in principle. Freud knew that any attempt to bring together psychoanalysis and neurology would be premature (although he himself made a last attempt at this in his 1895 "Project," which he left unpublished in his lifetime).

Neurology itself had to evolve, from a mechanical science that thought in terms of fixed "functions" and "centers," a sort of successor to phrenology, through much more sophisticated clinical approaches and deeper understandings, to a more dynamic analysis of neurological difficulties in terms of functional systems, often distributed widely through the brain and in continual interaction with each other. Such an approach was pioneered by A. R. Luria in the Soviet Union. But neuropsychology, as this approach came to be called, only got going during the Second World War, so, sadly, Freud never saw it, never knew how Luria had lifted clinical neurology to an entirely new level, a level perhaps complementary to that of psychoanalysis.

Indeed, Luria himself, as Solms has pointed out, was intensely interested in psychoanalysis as a young man and explored it quite deeply. But then, with the deepening intolerance of the 1930s, the very name of Freud became anathema in the Soviet Union, and it would have been impolitic, even suicidal, for Luria to continue. And yet Luria's first big book, his *Traumatic Aphasia*, published in 1947, had deep (if unacknowledged) indebtedness to Freud's *On Aphasia* of more than fifty years earlier.

Perhaps it could only be decades later that someone like Solms, equally trained in both neuroscience and psychoanalysis, equally drawn to both, could dream of conjoining Freud and Luria, conjoining the insights and approaches of neuropsychology and psychoanalysis, to aim at a science richer than either, a science that Solms sometimes calls "neuro-psychoanalysis" and sometimes "depth neuropsychology."

For classical neuropsychology, in a sense, only touches the surface of the mind—the surface character of perception, memory, language, thought, emotion, consciousness, personality, identity—and this is because of its objective and test-oriented approach. The appreciation of deeper determinants (which will be active in patients with neuropsychological problems no less than in the rest of us) requires the establishment of a genuine relationship between doctor and patient, a transference, the examination of resistances, and a patient attention to all that is said or not said, shown or concealed, and the use of free association to allow the mind maximum spontaneity.

Solms's approach, then, is a double one: to make the most detailed neuropsychological examination of patients with brain damage and then to submit them to a model psychoanalysis, and, in so doing, hopefully, to both deepen neuropsychology and ground psychoanalysis—to bring the mechanisms of the brain and the inner world of the patient together.

In addition to these essentially clinical approaches, there has been the development of wonderful forms of brain imaging in the past twenty years, making possible detailed studies of the brain's functional anatomy and metabolism, and also of experimental neuroscientific approaches to the mechanisms of emotion, attention, cognition, and consciousness. Thus the time for a synthesis draws closer and closer.

In all areas of science, but perhaps especially in biology and medicine, where the individuality of the organism and the particularities of life are all-important, one needs to have two sorts of book—clinical studies or case studies, in which neuro- or psychoanalysis are taken as far as they can go; and books organized according to concept and theme. Thus there are Freud's case histories—and there are his *Introductory Lectures*; there are Luria's case histories—and there are his expository books, such as *Higher Cortical Functions* (a highly technical book) and *The Working Brain* (a book designed to be accessible to anybody who is interested, who really wants to come to grips with neuropsychology). It is similar with Solms's books—we were recently given his brilliant *Clinical Studies in Neuro-Psychoanalysis* (coauthored with Karen Kaplan-Solms), and now we have this new thematic and systematic book, *The Brain and the Inner World*, written with Oliver Turnbull and aimed, as they say in their preface, at the nonspecialist reader.

There are some neurological or neuropsychological syndromes that seem easily compatible with psychoanalytic or metapsychological concepts—thus, large frontal-lobe lesions (such as that which affected Phineas Gage, after a tamping iron was thrust through his brain) can give rise to an impulsive, thoughtless, conscienceless state, sometimes called a psychopathy or pseudopsychopathy—and in such patients it is very much as if their superego functions have been extinguished (though not just these). The driven, violently appetitive states seen with some hypothalamic and periventricular lesions (or the absence of basic drives if there is massive destruction in these areas) suggest that it is in these primitive areas of central gray matter that much of the basis of the id resides. There are almost incredible states that may go with massive right-hemisphere lesions, where half the body may be neglected, disowned, or attributed preposterously to someone else—and such syndromes (it has seemed to neuroscientists like V. S. Ramachandran as well as to Solms) may involve a form of repression, and not just neural disconnection.

I am not wholly convinced myself that "repression" is an adequate term here, for one sees the most fantastical or delusional defenses against a catastrophic dissolution of the body-ego. Some people with Tourette's syndrome show explosive associative speed and license. When I was seeing a patient with such "phantasmagoric" Tourette's, I found Freud's *Interpretation of Dreams* and *Jokes and Their Relation to the Unconscious* to be indispensable, no less so than Luria's *The Mind of a Mnemonist*.

While it is unclear how far the correlation of neuropsychology and psychoanalysis can go, Solms has provided a brilliant and intriguing start to the process, partly by analogizing and theorizing, but most convincingly through beautifully studied clinical examples. Many people since Freud have explored the interpretation of dreams, the psychoanalytic approach, but who before Solms studied the neuropsychology of dreams, the ways in which their imagery, their style, even their existence could be transformed by damage to different areas of the brain? Many people, since Luria, have studied the neuropsychology of aphasias, parietal syndromes, right-hemisphere syndromes, frontal-lobe syndromes, and so forth. But who before Solms had explored them in psychoanalytic terms too, showing how certain psychoanalytic and metapsychological concepts had equally to be called upon? It is this which constitutes the heart of the neuro-psychoanalytical enterprise, the synthesis toward which Solms is reaching.

In his clinical studies, and now in this new book with Oliver Turnbull, Solms gives us not just scattered, suggestive examples, but detailed and systematic and scrupulously argued neuro-psychoanalytic studies of all his patients. In *The Brain and the Inner World*, after a lucid exposition of the anatomy and physiology of the brain, the authors summarize a vast amount of current neuroscientific work; in particular, they emphasize the pioneer neuroscientific work of Antonio Damasio and Jaak Panksepp, both of whom made major presentations at the first international conference on neuro-psychoanalysis, which Solms organized in 2000. *The Brain and the Inner World* covers the entire range, from emotion to motivation to memory and phantasy, dreams and hallucinations, words and things, the differential and complementary functions of the left and right hemispheres, a possible basis for the analytic "talking cure," the nature of unconscious and preconscious processes, and the very basis of subjectivity, consciousness, and self. Neuro-psychoanalysis, one feels, is spreading its wings, but always remaining, as it must, firmly grounded on the demonstrable and the testable.

One wonders how far this double approach will go, and what new regions it may be tempted to embrace. What goes on in the creative mind/brain? What is the basis of the Kierkegaardian categories: the esthetic, the ethical, the comic, the religious? Will psychoanalysis and neuroanalysis, separately or conjoined, provide an understanding of these fundamental human states? It is far too early to say. But it is clear that Solms and his colleagues are making a brilliant, determined, scrupulous, and (one wants to say) tactful endeavor to approach, in a new way, the oldest question of all—the mysterious relation of body and mind.

PREFACE

The "inner world" of the mind (*being* a mind and *living* a life) was, in the past, the traditional preserve of psychoanalysis and related disciplines, and it was therefore placed at the margins of natural science. This situation arose largely because neuroscientists did not consider subjective mental states (like consciousness, emotion, dreaming) to be suitable topics for serious brain research. However, in recent years—following the demise of behaviorism, the advent of functional brain-imaging technology, and the appearance of a molecular neurobiology—these topics have suddenly emerged from the shadows and assumed center stage in many leading neuroscientific laboratories around the world. Not surprisingly, this has produced an explosion of new insights into the natural laws that govern our inner life.

This book takes the nonspecialist reader on a guided tour of these exciting new discoveries, pointing out along the way how old psychodynamic concepts are being forged into a new scientific framework for understanding subjective experience, in health and disease.

Chapter 1 opens with a "*beginner's guide to the brain*" to familiarize readers with the basic neuroscientific terms and concepts that are required to navigate this field (and therefore the rest of this book). Chapter 2 introduces the *mind* into this neuroscientific matrix and asks a surprisingly difficult question: What, exactly, *is* "the mind"? We confront an ancient mystery: How does our immaterial consciousness—our very sense of existence and identity—emerge from the cell assemblies and other base processes of the brain, whose cells and processes are not fundamentally different from those of other bodily organs? Chapter 3, on the topic of *consciousness*, transforms this age-old *philosophical* problem into a *scientific* one: What, exactly, are the neural mechanisms that generate our awareness of our selves interacting with objects? We discover that these mechanisms are buried deep within the brain, that they are bound up with our most basic biological needs, and that they are inextricable from the brain mechanisms of *emotion*. Chapter 4 reviews the basic *emotional* mechanisms of the brain. The primal value systems that motivate all human behavior are identified. We learn that these core systems are deeply rooted in our evolutionary past, and that we humans share many of our most fundamental cares and concerns with other, "lower" animals. Chapter 5, on *memory*, describes how these inherited mechanisms are modified and individualized during development, and how our personal experiences are organized into predetermined categories of knowledge and behavior (some conscious and some unconscious) that shape our everyday lives. Chapter 6 brings together our understanding of the topics of the previous three chapters—consciousness, emotion, and memory—to help us unravel the secrets of *dreams*. After decades of "shrinking in horror at so impenetrable a problem" (as one eminent neuroscientist put it), we are finally beginning to understand the mechanism and meaning of dreams. Chapter 7 addresses another previously insoluble problem: the *nature–nurture conundrum*. To what extent are the trajectories of our lives predetermined by our genes? We learn how modern neurobiology answers this problem (using the topic of *sexual difference* as a paradigmatic example). Chapter 8 describes the functional differences between the *left and right hemispheres* and debunks some popular notions in the process (for example, the claim that the right hemisphere is the seat of the Freudian "Unconscious"). We consider how such speculations might, in the future, be tested scientifically. This raises the question of whether we are now in a position to translate Freudian theory into a series of testable hypotheses about the functional organization of the brain. Chapter 9 summarizes our arguments, attempts to draw the main strands together, and asks the integrative question: What is *the self* in neurobiological terms? And what, in neurobiological terms, might psychotherapists be doing when they *treat* a disordered "self"? Chapter 10 continues into this uncharted territory and concludes our journey by wondering whether it might finally be possible to bring the subject matter of psychoanalysis into the realm of natural science. What still needs to be done before this goal can be successfully achieved? We are introduced to the fledgling interdiscipline of *neuro-psychoanalysis*, which is now attempting, in the words of the latest Nobel laureate in medicine and physiology, to forge a "new intellectual framework for psychiatry" in the twenty-first century.

* * *

Acknowledgments are due, above all, to Maxine Skudowitz and Judith Brooke, and to Paula Barkay (the coordinator of the Anna Freud Centre lecture series, upon which this book is based). We are very grateful to those colleagues who read parts of this book with a critical eye, especially Jaak Panksepp and Derek Nikolinakos. We are also greatly indebted to Erica Johanson, our editor at Other Press, and to Klara and Eric King, our editors at Communication Crafts.

CHAPTER 1

INTRODUCTION TO BASIC CONCEPTS

This is very much a beginner's guide to the brain. It makes virtually no assumptions about previous knowledge of neuroscience, and there is no intention to dazzle the reader with exotic facts and stunning photographs. The aim is rather simple: to familiarize nonspecialists with the basic facts of how the brain "produces" our subjective mental life (as far as we understand these facts today).

To this end, each chapter provides an overview of the neurobiology of a particular aspect of the mind. The focus is on aspects that would traditionally have been the preserve of psychoanalysts rather than neuroscientists. In the past century, there was an unfortunate division between the subject matter of neuropsychology and the lived reality of the mind. This once prompted the neurologist Oliver Sacks to write that "neuropsychology is admirable, but it excludes the psyche"![1] Happily,

[1] Sacks elaborated: "Neuropsychology, like classical neurology, aims to be entirely objective, and its great power, its advances, come from just this. But a living creature, and especially a human being, is first and last . . . a subject, not an object. It is precisely the subject, the living 'I', which is excluded [from neuropsychology]" (Sacks, 1984, p. 164).

that situation has now changed. The really interesting things about psychology, such as consciousness, emotions, and dreams—topics from which neuropsychologists "shrank in horror" (Zeki, 1993, p. 343) less than a decade ago—are finally coming into the ambit of neuroscience. Readers of this book will learn what is known today about the neurobiology of *these* mental functions—about the "inner world" of the mind.

An example of personality change following brain injury

The following celebrated case illustrates why the inner world of the mind *should* be of interest to brain scientists.

In the 1840s, an unfortunate man by the name of Phineas Gage was laying railway tracks in the midwestern United States. He was pressing down a charge of dynamite into a rock formation, using a tamping rod, when the charge suddenly exploded. This caused the tamping rod to shoot through his head, from underneath his cheekbone into the frontal lobe of his brain and out through the top of his skull. Partly because the rod passed through so rapidly, probably cauterizing the tissue on its way, the damage to Gage's brain was not very widespread (Figure 1.1); only a relatively small area of frontal tissue was affected (for a

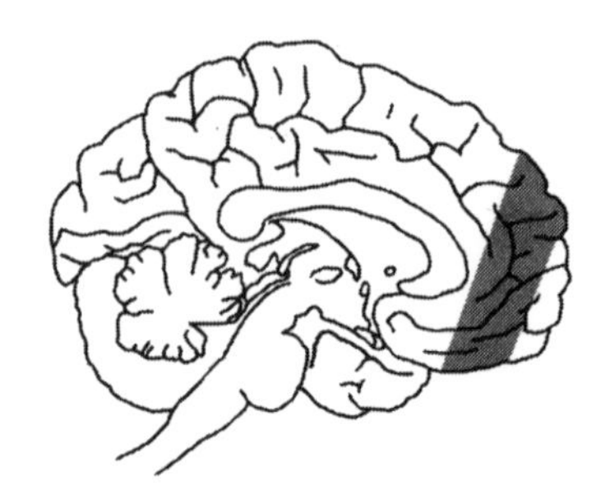

FIGURE 1.1
Phineas Gage's lesion

precise description see Damasio et al., 1994). Gage did not even lose consciousness, and he made a rapid physical recovery.

His physician, however, reported some interesting changes when he published the case in a local medical journal a few years after the incident. Dr. Harlow noted that, despite the good physical recovery and the relatively small extent of the brain injury, his patient was radically changed as a person; his *personality* was changed. Before the accident Gage had been the foreman of his team, a position of some responsibility; he was regarded as of reliable character and was highly valued by his employers. However, this is what Harlow said about Gage's condition after the accident:

> His physical health is good, and I am inclined to say that he has recovered . . . [but] the equilibrium or balance, so to speak, between his intellectual faculties and animal propensities, seems to have been destroyed. He is fitful, irreverent, indulging at times in the grossest profanity (which was not previously his custom), manifesting but little deference for his fellows, impatient of restraint or advice when it conflicts with his desires, at times pertinaciously obstinate, yet capricious and vacillating, devising many plans of future operation, which are no sooner arranged than they are abandoned. . . . In this regard his mind was radically changed, so decidedly that his friends and acquaintances said that he was "no longer Gage." [Harlow, 1868, p. 327]

Disregarding the quaint language, the message of this nineteenth-century physician's description still comes through clearly: as a result of his brain damage, Gage was "no longer Gage." The inescapable conclusion is that Gage's personality—his very identity—was somehow dependent upon the few cubic centimeters of brain tissue that were damaged in his accident. Today we know, from observing countless similar cases, that damage to that area of tissue almost always produces the very same type of personality change that it did in Gage's case. There is some variability, depending above all on the individual's personality before the injury, but these patients are *typically* fitful and irreverent, show little deference for others, are impatient of advice if it conflicts with their desires, and so forth. These are some of the cardinal features of what is now known as the "frontal-lobe personality."[2]

In our clinical work as neuropsychologists we have met literally hundreds of Phineas Gages, all with damage to the same part of the brain. This is a fact of obvious importance for anyone with an interest in personality. It suggests that there is a predictable relationship between specific brain events and specific aspects of *who we are*. If any one of us were to suffer the same lesion in that specific area, we would be changed in much the same way that Gage was, and we, too, would no longer be our former selves. This is the basis of our view that anyone with a serious interest in the inner life of the mind should also be interested in the brain, and vice versa.

TWO APPROACHES TO THE SCIENCE OF THE MIND

The mental life of real human beings is the traditional subject matter of psychoanalysis. We have said that it has recently become a legitimate subject matter for neuroscience too. In other words, we now have two disciplines (perhaps better described as two loose *groups* of disciplines) studying the same thing. But they approach this shared subject matter from completely different points of view.

[2]Anderson et al. (1999) recently observed that the "social emotions" fail to develop in children whose brains are damaged in this part of the frontal lobes, and Raine et al. (2000) noted that frontal-lobe volume is reduced in psychopaths.

The "subjective" approach to mental science (psychoanalysis) split off from the "objective" approach (the neurosciences) just over a hundred years ago. Freud's *Studies on Hysteria* (1895d) or his *The Interpretation of Dreams* (1900a) provide useful milestones in this divergence. Since then, each approach has developed along its own path. The original reasons for the split were complex (see Kaplan-Solms & Solms, 2000; Solms & Saling, 1986; see also chapter 10). Mainly it was a matter of expedience. It was not possible to learn anything useful about the mind—the *real* mind, in Oliver Sacks's sense—using the neuroscientific methods that were available at that time. Neuroscience could not (at that time) penetrate the mysteries of personality, motivation, emotion—the things that make us who we are—and it therefore seemed to Sigmund Freud that the most useful way to study, understand, and treat the disorders of the human subject was from a purely psychological perspective.

We do not wish to be excessively optimistic, but the reason that a book such as this one can be written today is because that situation has changed. We have powerful new methods and technologies in neuroscience that are yielding previously undreamed-of knowledge about the physiological underpinnings of the "inner world." In short, neuroscience has caught up with—many would say overtaken—psychoanalysis as a science of the human subject, and today it is possible to learn some very important and valuable things about inner experience by studying the *physical organ* that was damaged in the case of Phineas Gage.

RECONCILING THE TWO APPROACHES

It is essential for us to find some way of bridging this historical divide, and perhaps healing the rift, between these two different approaches to mental science. Neuroscientists—who are grappling with the complexities of human subjectivity for the first time—have much to learn from a century of psychoanalytic inquiry (see Kandel, 1998, 1999). Psychotherapists, for their part, have an opportunity to benefit from the enormous empirical advances in the neurosciences and, as a result, to make progress in their own disciplines, where scientific progress has become frustratingly slow. Psychoanalysis today is associated with bitter rivalry between opposing camps that apparently have no valid means of deciding between their conflicting standpoints on various theoretical matters. One solution might be to find links between the disputed theoretical concepts of psychoanalysis and those of the neurosciences.

This seems to be an appropriate way to proceed, but it is quite difficult to put into effect. There are a number of things that have to be done for us to be able to bridge the gulf that separates these two approaches. Each side has (for various reasons) regarded the other with suspicion and disdain for over a hundred years. Typically, neuroscientists have regarded psychoanalysis and related disciplines as "unscientific" (how can a science of subjectivity be objective?). Psychotherapists, for their part, have regarded the neurosciences (including biological psychiatry) as simplistic, to the extent of excluding the psyche. These attitudes have developed for good reasons, and they will not be overcome easily or quickly.

In addition, there are serious scientific problems to grapple with. How can we link these disciplines in a methodologically valid way? To take a concrete question, how do we set about identifying the neurological basis of something like, say, "repression"? How does one go about testing experimentally, from the neurobiological point of view, whether such a thing as repression even exists? Repression—if it exists—is a complicated, elusive, fleeting phenomenon. It is far from easy to capture such things in physiological terms.

If such problems are to be overcome effectively, a good deal of the effort required would have to be put in by members of *both* of the approaches working together. To do this, we would have to have interdisciplinary dialogues and research about topics of common interest. We would need to collaborate on clinical material and work together on the same cases, or on examples of the same disorders, to learn from each other's approaches. But first of all, before we can realistically combine them, we need to *learn* about each other's different perspectives.

This text goes under that "educational" heading. Our main aim in this book is to impart to readers raised on the language of psychoanalysis something about contemporary neuroscience, and in particular something about what contemporary neuroscience has to say about some topics of general interest. In doing so (and especially in the later chapters), we also hope to convey something of how neuroscience relates to what psychotherapists know and do, and how we might begin to make links between these two approaches to the mind.

What readers will *not* acquire from this book is a neuroscientific perspective on *particular psychopathologies*—such as neuropsychiatric perspectives on attention-deficit disorder, obsessive-compulsive disorder, tic disorders, panic attacks, and so on. These are very complicated topics in their own right and are beyond the scope of this general introduction to the field. We plan to write another book dealing with them in the not-too-distant future. One must become familiar with the field as a whole and know some elementary things about the brain in general and its basic mental functions before these more complex problems can be tackled. Fortunately, the neuroscience of these functions—consciousness, emotion, memory, and so on—involves some very interesting topics. But before we can delve into them, we first need to get some *really* basic facts out of the way.

ELEMENTARY BRAIN ANATOMY AND PHYSIOLOGY

We begin our task with an introduction to functional brain anatomy and physiology. Anatomy and physiology are not glamorous subjects—and a complete knowledge of them requires careful and intensive study. But they provide the very bedrock of the subject matter of this book. This section does not deal with them at the level of detail covered in a medical-school curriculum. We cover only those basic concepts that readers absolutely must be familiar with to easily understand the material discussed in the subsequent chapters. Readers who have a background in brain anatomy and physiology, perhaps from undergraduate medical or psychology courses, may wish to pass over the next few sections. Note, however, that a number of very important concepts are introduced in the section on "The Internal and External World" and again in that on "The Internal World."

It is easy to overlook the fact that the brain is, after all, *just an organ*. It is an organ like the liver or the spleen or the stomach. Like these other organs of the body, it is made of *cells*. These cells are connected together to form a piece of tissue with a certain characteristic texture and shape, and so the brains of all of us look roughly the same. And yet, there is something almost miraculously special about this organ: it is the organ of the *mind*—indeed, of our very *selves*, as Gage's case amply demonstrates.

Despite this unique property of the brain, its cells are not fundamentally different from the cells of other bodily organs. What is the prototypical nerve cell? It consists of three basic parts (Figure 1.2). The first, the **cell body**, contains essentially the same things found in cells in other organs—namely, the things that govern its basic metabolism. There are two types of appendages to this cell body, one of which is known as the **dendrites**, the other as the **axon**; in our prototypical nerve cell,

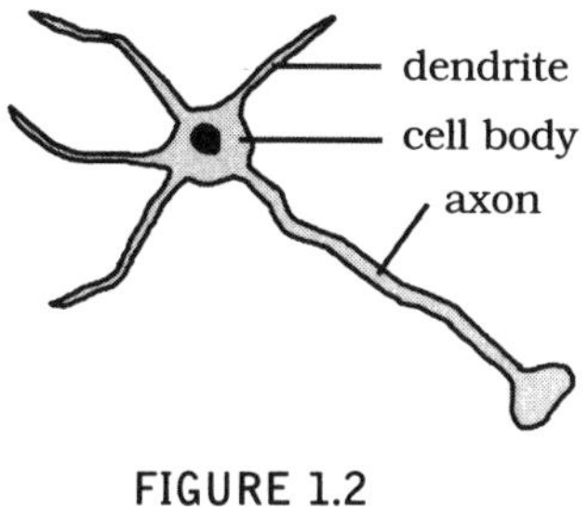

FIGURE 1.2
A nerve cell

there are many dendrites but only a single axon. Together, these three components form the typical structure of a brain cell—a **neuron**. Neurons (in conjunction with some supporting cells called *glia*) are all that the nervous system is made of—billions and billions of cells, connected up with one another.

This interconnection takes place as follows: The axon of one neuron links up with a dendrite of another neuron, whose axon in turn links with a dendrite of another neuron, and so on (Figure 1.3); multiple interconnections can occur, as each dendrite on a neuron can accept many axon terminals. At the place

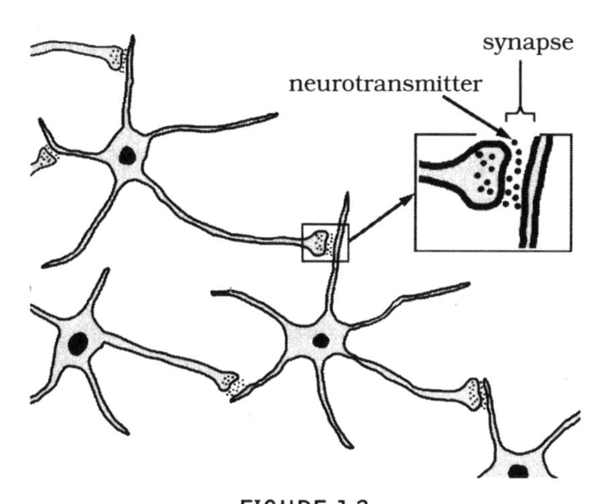

FIGURE 1.3
Nerve cells in series

where two cells link up—between the axon of one cell and a dendrite of the other—there is a minute gap, called a **synapse**. Over the synaptic gap, small chemical molecules pass from one neuron to the next; these molecules are called **neurotransmitters**. This transmission of chemicals is the principal means of communication between the cells of the brain. Different cells located in different brain regions use different types of neurotransmitters, some of which are introduced later in this chapter.

These five concepts—cell body, dendrite, axon, synapse, neurotransmitter—are all that one really needs to know about neurons for the purposes of this book.

What is it, then, that makes this organ so unique—how is it that these interconnected cells produce something as miraculous as our awareness of being in the world? How can it be that the physiological activity of these cells, comprising this lump of tissue, produces something so utterly unlike anything that any other organ produces—indeed, so utterly unlike anything else in the physical universe? In the next two chapters, this question is discussed in more depth.

Although the elementary properties of neural tissue obviously do not explain how or why the brain produces subjective awareness, there are two features about it that are quite unusual. These features are not fundamental, but they do distinguish the cells of the brain from those of most other bodily organs. The first distinguishing feature of neurons is the nature of the links between them: the *synapses* mediated by *neurotransmitters*. This linkage permits the passing of "information" from one cell to another. The principle of information transfer is not unique to nerve cells (other cells also interact with each other in various ways), but the *dedicated* function of communication between nerve cells is an important distinguishing feature.

The second outstanding feature of brain tissue is that, while the basic plan of the brain's organization is, as it were, predetermined by our genes (see chapter 7), the overall plan is dramatically modified by *environmental influences* during life. The brain comes into the world with innumerable *potential* patterns of detailed organization, as reflected in the infinite combinations through which its cells *could* connect up with each other. The precise way that they *do* connect up, in each and every one of us, is largely determined by the idiosyncratic environment in which each brain finds itself. In other words, the way our neurons connect up with each other depends on what *happens* to us. Modern neuroscience is becoming increasingly aware of the role played in brain development by experience, learning, and the quality of the facilitating environment—and not only during childhood (see chapters 5 and 7). In short, the fine organization of the brain is literally *sculpted* by the environment in which it finds itself—far more so than any other organ in the body, and over much longer periods of time.

At the level of neural tissue, then, these two features—the capacity for information transfer and that for learning—are what most distinguishes the brain from other organs. These capacities are present far more potently in brain tissue than in any other tissues of the body.

Gray and white matter

We have said already that neurons are connected to one another in their billions. Now, building on this, we must add that the cell bodies tend to *group* together, rather like debris on the surface of an expanse of water. When cell bodies clump together like this, the resultant tissue appears somewhat *grayish.* The stringy connections between the gray tissues, formed principally by the axons interconnecting the cell bodies, appear *white* by contrast (mainly because axons are surrounded by a sheath of fatty tissue, and fat has a white appearance). This is the basis of the famous distinction between **gray matter** and **white matter**

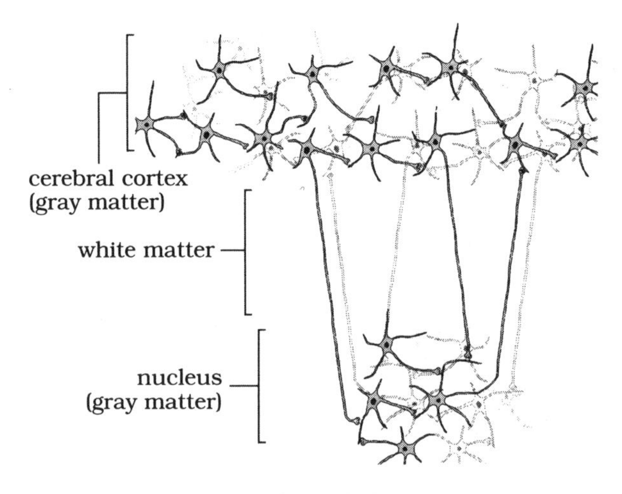

FIGURE 1.4
Gray and white matter

(Figure 1.4). Collections of cell bodies are gray; the fiber connections between them are white.

The cell bodies forming the gray matter group together in one of two ways—either as *nuclei* or in *layers* (see Figure 1.4). The nuclei are simply balls of cell bodies, lumped together. The layers are more complicated. They are formed when the cell bodies line up in rows. The resultant sheets of cells are typically found on the outer surface of the brain—and form its **cortex** ("cortex" means outer layer). There is a shortage of space in the human cranium, because the amount of cortex has expanded dramatically in recent evolution; so the brain saves space by folding the layers in upon themselves, in a wavelike pattern. This is what gives the outer surface of the brain its well-known convoluted appearance. The *nuclei* lie deeper within the brain, underneath these layers of cortex—and the white matter is located between the two. The white matter—principally axons—thereby connects the cell bodies of the nuclei and cortical layers with one another. The precise anatomy of the resultant systems is enormously complex, but these basic principles are easier to understand.

Brainstem and forebrain

A further basic division of the brain, which will be referred to repeatedly in the following chapters, is that between the brainstem and the forebrain (Figure 1.5). This is a distinction of great importance for understanding some of the psychological functions discussed later in this book. These two structures are, in turn, intricately subdivided. There are innumerable terms for the various regions within them—often (and quite confusingly) more than one term for the same structure. The terms we introduce now form the standard (or most widely accepted) terminology. It is not necessary for the reader to *remember* all these terms to follow the later chapters; it will be more practical to refer back to this chapter (using the index and the words emphasized in boldface type) for reorientation whenever necessary.

The **brainstem** is a direct extension of the **spinal cord** into the skull, and it is phylogenetically (i.e., in evolutionary terms) the most ancient part of the brain. In this book, we are more concerned with the nuclei *inside* the brainstem than with its outer surface. Accordingly, the best way for us to depict it is by slicing the brain down the midline to produce a medial view, as we did for Figures 1.1 and 1.5, showing the inner surface of one

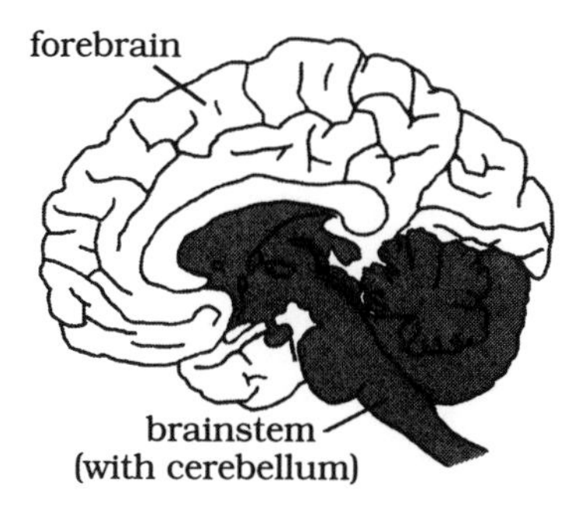

FIGURE 1.5
Brainstem and forebrain

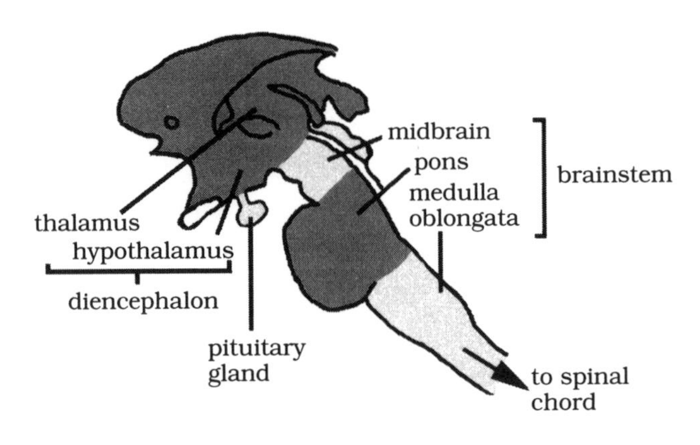

FIGURE 1.6
The brainstem

half (Figure 1.6). The lowest portion of the brainstem, the part immediately adjoining the spinal cord, is the **medulla oblongata** (Latin for "oblong core")—a structure that has little to do with what is traditionally conceived of as "the mind" (the medulla contains nuclei that govern heartbeat, breathing, etc.). Above the medulla oblongata is the **pons** (Latin for "bridge"). Hanging behind the pons is the **cerebellum** ("little brain"; see Figure 1.5). The top of the brainstem proper is the **midbrain**. Above this region are structures that are not technically part of the brainstem (opinions on this point have changed over the years), but they are very closely connected in functional terms to the medulla oblongata, pons, and midbrain. These structures are referred to as the **diencephalon**. There is no generally accepted English word for this region of the brain, although at one time it was called the "twixt-brain," which conveyed the essential fact that it lies between the brainstem and the forebrain. There are two main parts to the diencephalon. The largest, the upper portion, is the **thalamus**. Below the thalamus lies the **hypothalamus**, which is directly connected to the **pituitary gland** (Figure 1.6). All of these brainstem and diencephalic structures contain nuclei that are connected to one another (and to the forebrain structures described next) in intricate patterns. The details need not concern us yet, but some important ones will be discussed later.

The **forebrain** is phylogenetically younger than the brainstem. It consists principally of the two great **cerebral hemispheres** that fill the vault of the cranium. The outer surface of these hemispheres is the cerebral cortex, made up of folded layers of gray matter, as described above. Within the cerebral hemispheres, and hidden from view, are various forebrain nuclei (described below). Figure 1.7 depicts two views of the cerebral hemispheres. On the left side of Figure 1.7 is another medial view of the inner (flat) surface. From this angle we can see the **corpus callosum**, which is a bridge of white matter that connects the two hemispheres with each other. The right-hand perspective on the hemispheres is a lateral (i.e. side) view, showing the outer (convex) surface.

Each hemisphere is divided into four lobes. From both views in Figure 1.7 we can identify these lobes. At the back of the head is the **occipital lobe**; in the center is the **parietal lobe** (situated above and slightly behind the ears); below and in front of the parietal lobe is the **temporal lobe** (at the temples); the remainder of the hemisphere is the large **frontal lobe**, which lies over

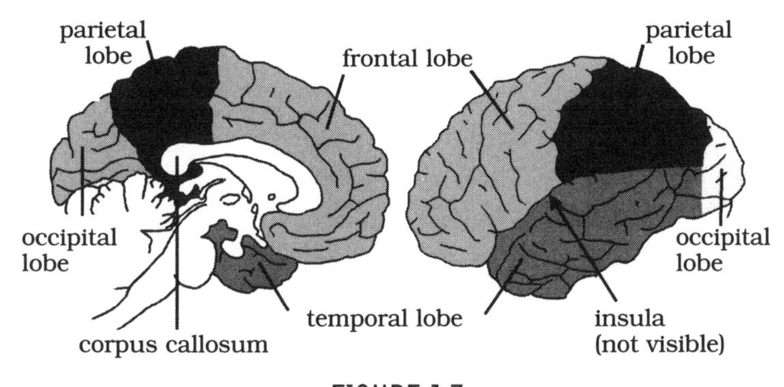

FIGURE 1.7
The forebrain

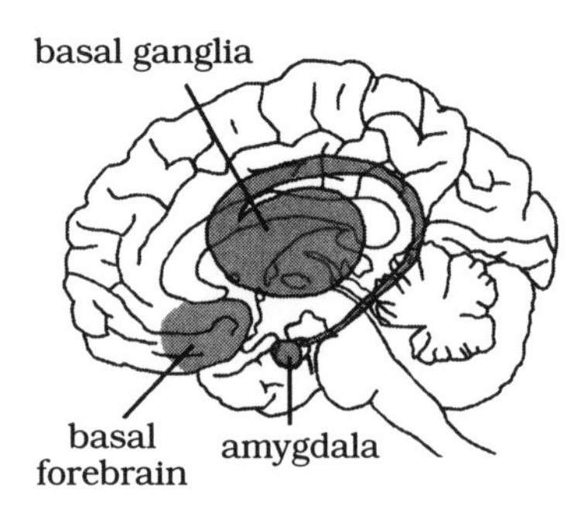

FIGURE 1.8
Deep forebrain structures

the eyes and is perhaps our greatest (and, in parts, uniquely human) phylogenetic acquisition. Buried between these lobes, if one pulls the temporal lobe down and lifts the frontal and parietal lobes up, lies a further region of cerebral cortex known as the **insula**.

Inside the cerebral hemispheres are the forebrain nuclei referred to above (Figure 1.8).[3] The most substantial such nuclei are the **basal ganglia**. Close to the basal ganglia, nestled within the lower half of the frontal lobe, are the **basal forebrain nuclei**. Behind them, inside the anterior (i.e. front) part of the temporal lobe, is the **amygdala** (Latin for "almond," which this group of nuclei resembles in shape).

The limbic system

The final anatomical term to which readers need to be introduced here is the **limbic system** (Figure 1.9). This term is frequently used as though it referred to an anatomical *structure*, but it is

[3] Some parts of these structures are in fact neither quite nuclei nor quite cortex—that is, they are nuclei with a layered structure (such transitional structures are called "corticoid").

really a *theoretical concept* about a group of structures that, many neuroscientists feel, are linked together in a functionally significant way. This group of structures features prominently in later chapters on emotion and memory (chapters 4 and 5). Because it is a theoretical concept rather than a concrete thing, different neuroscientists include different structures under the term "limbic system." It is therefore a rather vaguely defined entity (the very usefulness of which some neuroscientists question). However, more or less everyone includes the following structures in it. At its core is the **hypothalamus**. Around this core, and connected with it, the other limbic structures are arranged in a ringlike formation. Within the diencephalon, we include part of the **thalamus** (most theorists include the *anterior and dorsomedial nuclei* of the thalamus in the limbic system). Outside the diencephalon, in the temporal lobe, we include the **amygdala** and the **hippocampus,** together with a fiber pathway called the **fornix,** which courses under the corpus callosum as it links back to the diencephalon, where it joins the hippocampus to a small nucleus called the **mamillary body**. The hippocampus is not a nucleus but, rather, consists of a phylogenetically old kind of cortex, running along the inner surface of the temporal

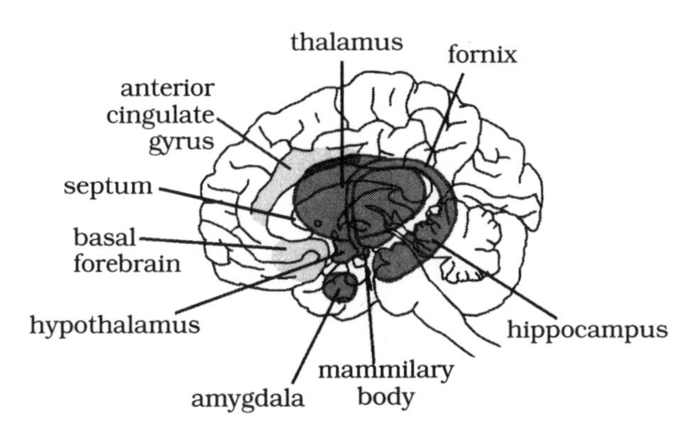

FIGURE 1.9
The "limbic system"

lobe. It is also strongly connected to the group of **basal forebrain nuclei**, including those embedded in the **septum**. Several of these structures too are connected to the **anterior cingulate gyrus**, which is therefore also usually included in the limbic system.

This highly interconnected set of brain structures, most of which lie deep within the brain, comprises the limbic system. There are many other structures that connect with these in complicated ways, some of which are also sometimes considered "limbic." However, these are not *core* components of the limbic system, and we shall only introduce them as we bump into them in later chapters.

This concludes the basic anatomical material required to navigate the chapters of this book with ease. For many readers, this section will have been the most difficult part of the entire book. The anatomical terms introduced here will be mentioned time and again, and repeated exposure (especially in the context of discussions of their psychological functions) will lead to much greater familiarity with these terms and the anatomical structures to which they refer.

THE INTERNAL AND EXTERNAL WORLDS

The brain is an organ, but it is not an *isolated* organ. It is connected in various ways with the other organs of the body. This vital fact about the brain and how it works is all too often overlooked (especially by people who like to think of the brain as something analogous to a computer).

At this point we must introduce a distinction that will run like a red thread through all the subsequent chapters of this book. In a nutshell, the brain is connected to two "worlds": the world *within* us, the internal milieu of the body; and the world *outside* us, the external environment. In a profound sense, *the principal task of the brain is to mediate this divide*—to mediate between the vital requirements of the internal milieu of the body (the vegetative functions) and the ever-changing world around us, which is the source of everything our bodies need but is indifferent to those needs (with the exception of our parents—especially during childhood—and other loved ones, who for that very reason occupy a special place in the mental economy).

The vegetative nervous system performs the task of keeping the body alive from moment to moment, controlling heart rate, breathing, digestion, temperature, and so forth. To perform these functions, the body requires, and actually *consumes*, material from the outside world—principally food, water, and oxygen. It also requires a suitable ambient temperature, as the organs of the body can only function within a very narrow temperature range. The same applies to sexual needs—though sexual "consummation" is necessary for the survival of our species as a whole rather than of each one of us individually. In short, to maintain and sustain the visceral jellyfish that we all have inside us, the internal world of the body has to interact in an appropriate way with the external world around us and make that world meet its needs, and it is the *brain* that manages this difficult task. When the external world no longer meets our many needs (when the brain is no longer able to regulate the inner functions of our bodies, by virtue of its interactions with the external world), we die—of hunger, thirst, suffocation, heatstroke, or one of the many other hazards that constantly threaten the integrity of the internal world of the body.

This point is obvious, and clearly irrevocable. *How*, then, does the brain perform these vital functions? In a broad sense, the answer to that question is what this whole book is about. To begin with, though, we can address a narrower question: How is the brain linked, *anatomically and physiologically*, with the inner and outer worlds of the body?

Perceiving and representing the external world

The brain is connected to the *outside* world in two main ways. The first is through the sensory apparatus (the organs of vision, hearing, etc.); the second is through the motor apparatus (the so-called musculoskeletal system). This is how we receive information *from* the world and how we act *on* the world.

There is no need to go into the precise details of sensorimotor functioning here, though the neurobiological basis of these mechanisms is well understood—in some respects to an extraordinary level of detail. The essential facts are that sensation is generated by specialized sensory receptors (in the eye, ear, etc.) that transform selected physical features of the environment into nerve impulses and send the resultant information to the brain. In the case of vision, cells in the retina send (most, but not all) visual information, via part of the thalamus, to the back of the occipital lobes. A similar arrangement applies to hearing, in which case (most, but not all) auditory information is transmitted (via a different part of the thalamus) to the superior (i.e. upper) surface of the temporal lobe. In the case of somatic sensation (touch, pain, etc.), the relevant information is sent from the surfaces and joints of the body to (mainly) the anterior part of the parietal lobe.

Somatic sensation (sensation arising from the skin, muscles, and joints of the body) is in fact quite a complex matter, and requires a little more discussion here. Many people refer to this as the sense of "touch." In fact, touch is part of a *group* of different sensory modalities that transmit several types of information from the surfaces of the body, of which tactile sensation is only one. There is also vibration sense, temperature sense, pain sense, and muscle- and joint-position sense. Each of these could be regarded as a sense modality in its own right, in that each is served by a specific type of receptor and projects separately to the brain. However, all these sense modalities send information to a roughly similar location in the brain, in the parietal lobe, which forms the basis of the body schema, and they are therefore grouped together as "somatic sensation."

It is important to be aware that the modality of somatic sensation carries only *part* of the information about the state of the body to the brain—that is, information about the *external* aspect of the body, the "musculoskeletal" part, which is in contact with the outer world. We need to know about pain and temperature in the outside world in order to act on it. Information about the *internal* world of the body, relating to the viscera, is not conveyed by sensors such as those for touch, pain, and so forth.[4] The internal world is discussed later in this chapter—for the moment we are discussing only how the brain is connected to the external world.

The two remaining sensory modalities—taste and smell—are "chemical" in nature. The precise anatomy and physiology of these systems need not trouble us. Taste is closely connected to somatic sensation in the tongue and is represented mainly in the cortex of the insula. Smell is connected to a range of structures inside the temporal lobe, including some parts of the limbic system.

That concludes our brief overview of the neurology of the five classical sensory modalities. For now, all that readers need to remember is that the three main sensory modalities are projected onto the three lobes at the back of the brain: the occipital lobe for vision, the temporal lobe for hearing, and the parietal lobe for somatic sensation.

[4] Some information about some internal organs is conveyed (often in a degraded form) via the somatosensory system—hence stomachache, etc.—but the great bulk of information about the state of our internal organs is conveyed via a different system, to be described later in this chapter.

Acting on the external world So far we have said nothing of the frontal lobes. Whereas the sensory organs are connected principally to the back half of the brain, the *motor* organs are connected principally to the front half (though they are also regulated by a range of other structures, mainly in the basal ganglia and cerebellum). The neurobiology of the action system is fairly well understood, though perhaps not as precisely as we understand the various perceptual modalities. In part, this is because the motor system acts on the basis of information derived from *all* the senses. (That is what the senses are *for*: they guide action.) For example, visual control of action usually operates in concert with feedback from the various somatosensory systems, which provide information about (say, initially) joint position. Once an object is grasped, the touch system provides feedback, which modulates the strength of the grip. Anyone who has tried to act on the world while an arm is numb after resting on it, or to speak after a dental anesthetic, will recognize the interaction of the systems involved. In addition, there is a range of action systems that operate in concert. We not only reach out to grasp objects, but in doing so we also move our eyes to search for the objects.

As an aside, it is worth reminding ourselves that these simple facts represent some 150 years of accumulated knowledge in neuroscience (here summarized crudely in a few lines). Before psychoanalysis was born, Freud contributed to this knowledge by tracing an aspect of how auditory information reaches the brain. He made an important contribution by describing the termination of the acoustic nerve in a nucleus of the brainstem, a location where auditory nerve impulses travel from the ear on their way to the cortex (Freud, 1886). This was the level of knowledge that Freud was able to glean from the neurosciences, and to a great extent this was the level of knowledge in neuroscience in general at the time that psychoanalysis was born. It is easy to imagine Freud thinking that far more needed to be known than how the sensory modalities and muscle systems are projected onto the surface of the brain if understanding was ever to be achieved of what governs the thoughts and feelings of a human being.

"Projection" and "association" cortex We have said already that visual information arrives in the occipital lobe, auditory information in the temporal lobe, and somatic sensation in the parietal lobe. Thus, there is a "visual," an "auditory," and a "tactile" part to the cerebral hemispheres. We have also said that "motor" impulses arise from the frontal lobe. However, only *part* of each lobe is dedicated to the control of these modalities; these parts are called the **projection** cortex (they are sometimes also called *primary* or *sensorimotor* areas). In these regions the nervous tracts derived from the various sensorimotor organs literally *project* the receptor and effector surfaces of those organs onto the cortex, forming tiny functional maps of the body all over the brain (see Figure 1.10). Sensory information (i.e., the pattern of stimuli impinging on the receptive surfaces) is, to some extent,

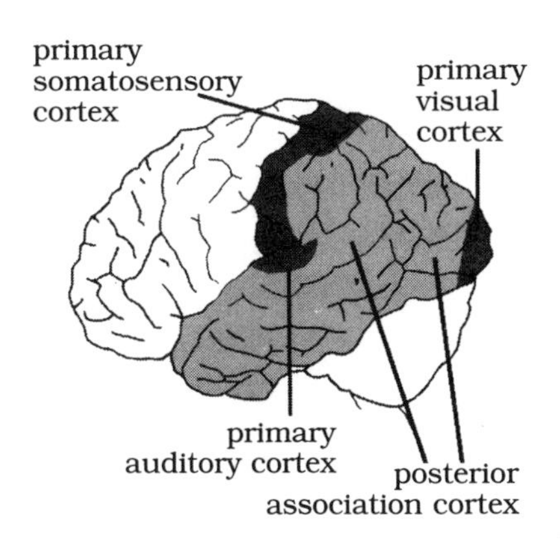

FIGURE 1.10
Primary sensory and association cortex

modified in the process of traveling to the cerebral cortex. However, the pattern of neural activity representing that information in the primary projection cortex still retains its original topographic organization with respect to the receptive field of the sense organ. Thus, for example, the lower-left corner of the retinae (representing the upper-right corner of the visual fields) is always mapped onto the lower half of the visual projection cortex of the left occipital lobe. Likewise, there are little maps of the entire surface of the outer body, representing the various submodalities of somatic sensation, projected onto the primary sensory cortex of the parietal lobe (lying upside down).

The primary projection cortex, which maps the receptive surfaces of our sense organs more or less directly onto the brain, occupies only a small portion of the hemispheres (Figure 1.10). Between the projection areas we have a range of complex brain areas, which are specialized for a variety of cognitive functions involved in the *further processing and storage* of incoming sensory information. In the case of vision (which is the best-understood modality), there are, for example, regions for processing location, color, and motion. At a higher level there are systems for object recognition, attention, and visuospatial manipulation—a cascade of extremely complex systems, the further details of which need not concern us here. These regions are often referred to by the catch-all phrase of the **association** cortex (see Figure 1.10). This term reflects the fact that the association cortex is involved in *integrating* the information derived from the different modality-specific (projection) regions. The further the information is integrated, the less modality-specific (and the less localized) the representational maps become. Thus, the object-recognition system in the association cortex can ultimately recognize a "dog" on the basis of either visual, auditory, or tactile information. This is achieved through the construction of neural *directories* that link all the relevant bits of information with one another.

Within the association cortex, therefore, memories are laid down, because a system designed to recognize the outside world must also store knowledge about that world. However, not *all* types of memory are stored in the association cortex; many other brain regions are involved in the totality of mental processes included under the term "memory" (see chapter 5). Continual exposure to a range of perceptual experiences allows for the development of well-established memories. Thus, there are many different ways in which you can experience a "dog"—not only through different sense modalities, but also through the ways that a dog can appear from different viewing perspectives, the different shapes and breeds of dog, the types of movement that dogs can and cannot produce, the places in which you are most likely to find dogs, and so forth. On the basis of millions of experiences, we gradually build up a reliable and stable picture of the outside world.

From what has been described so far, it should be clear why the posterior (i.e rear) half of the hemispheres of the brain is traditionally defined in neuropsychology as a functional unit for the *reception, analysis, and storage of information* (Luria, 1973). These functional processes constitute the very fabric of our experience of the outer world.

As discussed earlier, the *anterior* parts of the cerebral hemispheres are concerned with *motor* processes. Here, principles similar to those just described apply, in that there is both projection and association cortex (Figure 1.11)—though motor association cortex integrates *plans of action* rather than perceptions. The most anterior parts of the frontal lobes (which are confusingly called the *pre*frontal lobes) receive information about the state of the outside world from the posterior association areas described above. On this basis, a "likely" course of action is planned, preliminary to action itself (see chapter 9). The action itself is then implemented by the most posterior parts of the frontal lobes (near the middle of the brain) which form the

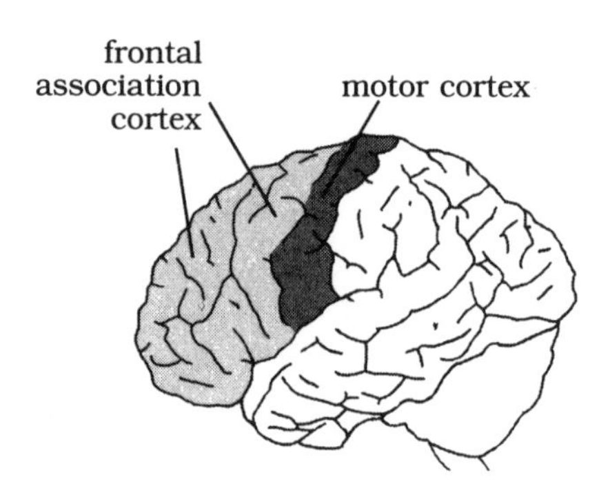

FIGURE 1.11
Frontal association and projection areas

primary initiation system for muscles of the limbs, trunk, and head. The action system is thus connected more or less directly to the outside world, in an analogous fashion to the way that information is derived *from* the outside world in the posterior projection cortex. Accordingly, the musculature of the body is mapped onto the primary motor cortex in a somatotopic fashion (i.e., there are little motor maps too).

It is important to remember that the action system always operates in concert with the perceptual systems (the primary function of which, as stated already, is to guide action). Moreover, we mentioned earlier that the prefrontal lobes receive information about the state of the outside world from the posterior association areas, and we should add that this always includes the state of the world *consequent upon the subject's previous action*. This constant *feedback* loop entails an extremely important self-monitoring function. Accordingly, whereas the posterior half of the forebrain is traditionally described as a functional unit for the reception, analysis, and storage of information, the anterior half is defined as the functional unit for the *programming, regulation, and verification of action* (Luria, 1973).

A few more words on the issue of self-monitoring are appropriate here. These cortical systems are not simple reflex arcs, where information enters and then is acted on rapidly and automatically. The prefrontal lobes offer the possibility of generating "potential" actions. Thus, in problem solving, the system can check to see whether the imagined solution matches the initial demands of the task (whether a particular approach will be "good enough") by *mentally* trying out actions, without endangering the body in completing the real event. In addition to this "feedforward" function, there is the regular feedback function mentioned already, monitoring what has just happened in the world and asking: "Has what I wanted to achieve occurred yet?" or "Has the situation changed now?" Such monitoring processes take place regularly when this system is functioning well, but they do not operate *constantly*. For example, the system is typically inactive when we are performing a routine task, so that we sometimes make "action slips," which represent the operation of automatic responses when the current situation actually required a modification of the system. (The link here with Freudian psychology is obvious.) The errors of judgment and problem solving that we discussed in the case of Phineas Gage were the result of massive damage to this system.

In summary, then (and as a broad generalization), various regions of the cerebral hemispheres participate in dealing with the outside world. The *posterior* regions *receive* information from the outside world and process it in accordance with previous experience, to establish what objects are of interest. The *anterior* systems *act* on the external world. In traditional faculty psychology, the operation of these systems would be classified under the headings of perception, memory, cognition, and action. They represent one entire class of mental life—interaction with the environment. Over most of the last century, neuroscience studied this aspect of the mind almost to the exclusion of any other, and tremendous progress was made. Until very recently, there was far less investigation of neuropsychological matters pertaining to the second aspect of reality—the influence of the *inner* world of the body on our mental life.

The internal world

We have said that the brain is interposed *between* two worlds—the external environment and the internal milieu of the body. The internal milieu refers to the world of respiration, digestion, blood pressure, temperature control, sexual reproduction, and the like. These organs are responsible for the body's survival, and in most cases loss of their functions would mean a rapid end to the life of the organism.

It may surprise some readers to hear that the operation of the viscera is of critical importance for understanding the "inner world" in the *psychological* sense too (i.e., the world of subjective experience). It is critically important because the operation of these systems forms the basis of our basic motivations or "drives" (as Freud called them),[5] and modifications in our drives are experienced, above all, as *emotions*. This issue is so important that an entire chapter of this book (chapter 4) is dedicated to it. Modifications in these internally directed brain systems also affect our level of *consciousness* in general (discussed in chapter 3). In fact, emotion and consciousness turn out to be inextricable, as we shall see here in chapters 3 and 4. For these reasons, the visceral component of the brain is traditionally defined as a functional unit for *modulating cortical tone and arousal* (Luria, 1973). However, as we explain in chapter 5, it is also critically involved in an aspect of autobiographical *memory* known as "episodic" memory. This aspect of memory, too, is inextricable from emotion and consciousness. Only a brief synopsis of these extremely interesting systems is provided here, as they are discussed in detail in chapters 3, 4, and 5.

---

[5]The term "drive" means different things to different people. Freud's definition was: "the psychical representative of the stimuli originating within the organism and reaching the mind, as a measure of the demand made upon the mind for work in consequence of its connection with the body" (Freud, 1915c, p. 122). This definition is consistent with the way we use the term in this book.

Briefly stated: Information travels up through the spinal cord (and in other ways, to be discussed later) from the interior of the body. This information reaches, in the first instance, the hypothalamus—which is the controlling mechanism (or "head ganglion") of the autonomic nervous system (the system that controls the self-regulating aspects of the body). The hypothalamus is intimately connected with the group of structures known as the limbic system, discussed above. One could say (using the language employed above in relation to external perception and action) that the functions of the internal milieu are "projected" onto the hypothalamus. The hypothalamus relays this information to a range of other structures throughout the limbic system and rest of the brain. In this way, the prevailing state of the body is linked with concurrent objects in the external world, and these links (which are of crucial importance for survival) are committed to memory. Following our analogy, the limbic system as a whole may be regarded as the "association" area for visceral information. The perception of visceral information is registered consciously as *feelings* of emotion and (via association) as reminiscences, in the form of: "I saw that, and it made me feel like this" (see chapter 5).

Alongside the "perceptual" aspect of this internally directed system, there is also a "motor" component. There are two classes of action performed by this system. One influences the visceral milieu itself (via secretory discharges, vasomotor changes, etc.). These influences are mediated by the autonomic nervous system. But the visceral brain influences *external* action too. External action is mediated by the motor systems already discussed above, but, unlike voluntary action, the visceral brain releases *stereotyped* motor patterns, executed under compulsive pressure. This is the basis of **instinctual** behaviors and the *expression of the emotions.* Unlike voluntary action, this type of motor activity is mediated primarily by the basal ganglia. However, information about the state of the internal milieu also reaches the prefrontal lobes—where it makes an important contribution to the calculations performed by the unit for the programming, regulation, and verification of action.

Executive control An extremely important aspect of prefrontal-lobe maturation involves the gradual development of *inhibitory* control over the stereotyped motor patterns released by the visceral systems of the brain. The developing prefrontal lobes also gain inhibitory control over emotionality and consciousness in general, thus providing the basis of directed thinking and attention, and so forth. When we say that the biological purpose of perception is to guide action, this applies no less to internally derived perceptual information than to that derived externally. For this reason, the prefrontal lobes are said to form a *superstructure* over the brain as a whole; this region of the brain governs our behavior (formulates and constantly monitors and modifies plans of action) on the basis of information derived from *both* the inner and the outer worlds (see chapter 9).

TWO SOURCES OF INFORMATION, SUMMARIZED

In summary, then, it is possible to develop a relatively simple picture of the entire system—a broad overview of how the brain as a whole operates. The brain is interposed between the outer and inner worlds of the body. From the external environment, information arrives through the sense organs and is directed to the posterior parts of the cerebral hemispheres. Information derived from each class of sensory receptor is projected onto the primary cortex specifically designed to deal with each modality, whereafter it is linked with other bits of information, mainly in the association regions in the posterior parts of the hemispheres. Integrated with traces of previous experience, this knowledge of the outside world is transmitted to the frontal association cortex, where it guides action programs. These programs are equally guided by influences derived from the internal world of the body. The internal information is registered, first and foremost, by the hypothalamus, and associated with other information in the limbic system, before it is conveyed to frontal cortex. This is the source of our inner motivation, which is intimately linked with personal memory, emotion, and consciousness. In this way, the prefrontal system is equipped to govern behavior, on the basis not only of prevailing external and internal conditions but also of previous experience.

This last statement carries implications, of course, for what would happen in cases where one of the systems influencing the prefrontal lobes is modified in such a way that the "balance" between them is disrupted. Someone might, for example, be too readily influenced by his or her drives and hence might act compulsively in the service of short-term goals, and in a way that is inappropriate from the viewpoint of the external world. Alternatively, disruptions to the internal source of information might lead to inertia, or to failure to modify behavior on the basis of emotional markers of previous experience. Later chapters discuss these issues in more detail. It should then become clear how abnormal behaviors can occur—whether through developmental abnormalities, brain damage, or distortions of the chemistry of the relevant brain systems. Armed with a knowledge of basic anatomy and physiology, the essential features of such changes are not too difficult to understand. So far, however, we have mainly discussed *anatomy*. Now we must turn briefly to the *physiological* principles that underpin the brain's functioning.

BASIC NEUROPHYSIOLOGY

The brain is made up of neurons, together with a range of nonnervous cells that act in support of neurons and help to maintain their survival. We said earlier that one of the unique properties of the living neuron is its capacity to transmit information. It does this by "firing." This term denotes the fact that every cell periodically transmits small quantities of neurotransmitter to its neighboring neurons. All cells in the body absorb and expel molecules. Neurons do this in a special way. Neurotransmitter molecules are expelled from the end of the axon of the neuron, into the small space separating it from the next cell, the synapse (see above). The neurotransmitter substance is then taken up by receptors on the dendrites of neurons on the other side of the synapse. This affects the second set of neurons by increasing or decreasing the chances that they will fire. Thus, neurons are in constant communication with each other through neurotransmitters. The communication is constant. Neurons always have a base ("resting") rate of firing; even when they are not specifically stimulated by other neurons, they fire at regular intervals. However, the action of other neurons, via their neurotransmitters, modifies the base firing rate—making each neuron fire more, or less, frequently than its resting rate.

There are two general types of neurotransmitter: **excitatory** and **inhibitory**. The excitatory type (the most common) *increases* firing rates—or, more precisely, it increases the chances that the next neuron will fire. It increases the *chances* of it firing, because we are actually dealing with aggregates of large numbers of neurons firing in concert. Each neuron does not connect to one further neuron and compel it to fire. Each neuron is influenced (via multiple neurotransmitters acting at multiple synapses) by dozens, even hundreds or thousands, of other neurons. Thus, the reception of an excitatory neurotransmitter increases the *chances* of the neuron firing. Similarly, an inhibitory neurotransmitter *decreases* the chances of that neuron firing. Because we are dealing with aggregates of neurons, it is the overall "average" outcome that will determine whether the neuron fires or not, or rather the *rate* at which it fires. To take a crude example, if 60% of a neuron's inputs are exciting it and 40% of them are inhibiting it, it is going to fire, but at a level not much above its base rate. If 90% are exciting it and 10% are inhibiting it, it is going to fire at a much faster rate. The complete mechanism of neurotransmission is more complex than we have described—for example, neurons are equipped with different synaptic receptors that receive, or "recognize," different neurotransmitters—but this preliminary account conveys the essentials in sufficient detail for our purposes.

So that is how neurons *work*. Again, it is worth noting that there is nothing mystical about these processes that "produce" the mind. They are just ordinary cellular processes. How they produce our beloved selves, with all the richness of our inner life, must involve something more than the simple facts of neurotransmission (see chapter 2). In the previous sections, we began to describe the pattern of connectivity between arrays of neurons—and how it enables them to integrate information about (and act upon) the outer and inner worlds. Now we must say something about the *physiological* principles that distinguish these two great divisions of the brain.

"Channel" versus "state" functions

The outer and inner sources of information can be distinguished not only on *anatomical* grounds, but on physiological grounds as well. The basic physiological division is embodied in the distinction that some neuroscientists draw between "**channel**" and "**state**" functions—the terms introduced by Mesulam (1998). His terminology is fairly idiosyncratic, but it denotes a relatively conservative concept, the physiological foundations of which are widely accepted. Other neuroscientists distinguish between the "contents" and "level" of consciousness—but these terms are less serviceable as they refer specifically to *consciousness* and thereby exclude the possibility of *unconscious* mental processes (see chapter 3).[6] Mesulam's distinction between the channel and state functions of the brain is perhaps roughly equivalent to the distinction that psychoanalysts draw between mental *representations* ("ideational traces") and mental *energies* ("quotas of affect").

Brain functions (principally, forebrain functions) dependent on information derived from the *external* world are primarily *channel*-dependent functions. This means that the information processed by these systems comes in *discrete bits* and is communicated via *distinct and specific pathways*. Information transmitted from a particular source within a channel-dependent system is not widely distributed to the brain as a whole but, rather, is targeted with great accuracy to other discrete regions. For example, when information arrives at a particular location on the retinae (say 30 degrees below the horizontal and 20 degrees left of the vertical meridian), it projects to a highly specific area of primary visual cortex which represents that precise location on the retinae (and therefore in the external visual field). The color aspects of this information then project to specific color areas, as do the motion aspects, and so on. In each case, a limited number of neurons directly "speak to" a limited number of other neurons some distance away, while *the vast bulk of the brain is completely unaffected by this interaction*. Thus, Region A connects to Region B, which connects to Region C. Regions L, M, and N, which also connect with each other, are never involved in the interaction between Regions A, B, and C (see Figure 1.12). Ring-fenced interaction of this kind occurs not only in the visual system, but in more or less all the externally directed functional systems of the brain.

[6]This distinction is similar to the distinction that Freud drew between mental "quality" and mental "quantity."

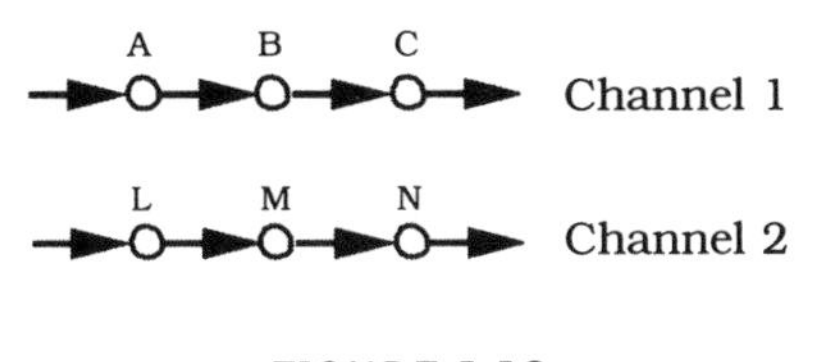

FIGURE 1.12
Channel functions

This type of interaction between neurons involves three main neurotransmitters. The principal excitatory neurotransmitters are **glutamate** and **aspartate**. The principal inhibitory neurotransmitter is **GABA** (*gamma-aminobutyric acid*). These are the most common neurotransmitters in the brain, and they dominate the activity of all channel functions.

The *internally* directed brain structures, described earlier as a unit for modulating cortical tone and arousal, operate in an entirely different way. Here, the means of communication is more gross and involves *widespread and global effects* that reflect changes in the *state* of the organism rather than in specific information-processing channels. The neurons of single brainstem nuclei in the state-dependent systems project onto extremely large numbers of other neurons—at ratios that are out of all proportion to the number of neurons in the source nuclei. The forebrain neurons thus affected are extremely widely distributed within the brain, so that a nucleus in the brainstem can influence neurons in all lobes of the forebrain simultaneously. In addition, forebrain neurons affected by one state-dependent nucleus can simultaneously be influenced by another one; in these systems there are no specific pathways (channels) but, rather, a number of overlapping "fields of influence." Figure 1.13 contrasts this mode of communication with that depicted in Figure 1.12. State 1 in Figure 1.13 affects both channels globally, whereas State 2 affects only part of both channels. The specific serial linkages between regions in the channel systems are thus replaced by overlapping and interacting *fields*. Even more

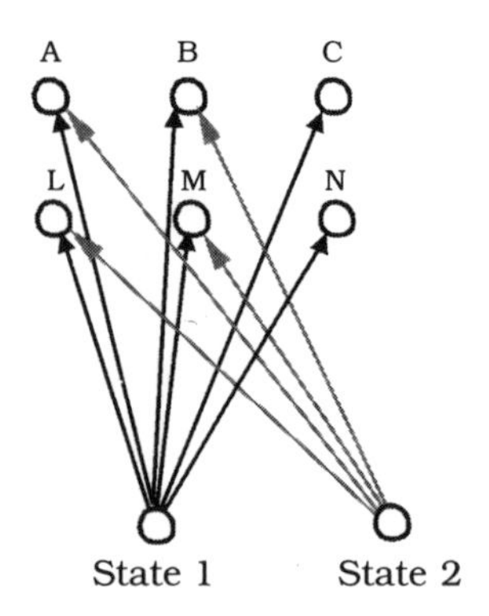

FIGURE 1.13
State functions

characteristic of the state-dependent systems is the fact (discussed later) that they are also open to influence by chemicals other than neurotransmitters, which link the brain *directly* with the visceral body.

BASIC NEUROCHEMISTRY AND PSYCHOPHARMACOLOGY

We said earlier that channel functions operate primarily with the classical neurotransmitters *glutamate*, *aspartate*, and *GABA*. The state functions too operate with these same neurotransmitters, but also with a number of others, such as *serotonin* and *dopamine*. The latter terms may be familiar to readers because psychopharmacologists work constantly with these neurotransmitters—and herein lies an interesting connection. It is no accident that the most familiar aspects of psychopharmacology deal with *these* chemicals, which convey the influence on the brain of the *internal* milieu—the "drives." What, then, are the chemicals that govern the internally directed systems?

The first is the neurotransmitter **acetylcholine** (ACh), which is employed by a good many neurons throughout the brain.

Neurons using this neurotransmitter are called *cholinergic* neurons, and two such systems are of specific interest (Figure 1.14). The first cholinergic system arises in the **mesopontine tegmentum** (part of the **reticular formation**, in the back half of the pons). These neurons project via the thalamus and influence the cortex in a fairly global way. Only the *cell bodies* of these acetylcholine-producing neurons are found in the brainstem structures of the pontine tegmentum. The *axons* of these cells extend into sites in the hypothalamus, thalamus, and cerebral cortex, where they release acetylcholine into synaptic spaces adjacent to other cells and modify the firing rates of those cells with ACh receptors. This arrangement—that is, neurotransmitter systems having narrow sites of origin (cell bodies clumped together in nuclei) and broad regions to which they project (via their axons)—applies to all the systems described in this section. A second (and very important) state-dependent system that employs ACh has its origin in the *basal forebrain nuclei*, which were mentioned earlier. This system, too, globally affects the firing rate of almost the entire cortex. (The cholinergic systems of the brain are discussed again in chapter 6, in relation to sleep and dreaming.)

The next important state-dependent neurotransmitter system has its origins in the **raphe nuclei** of the brainstem (see

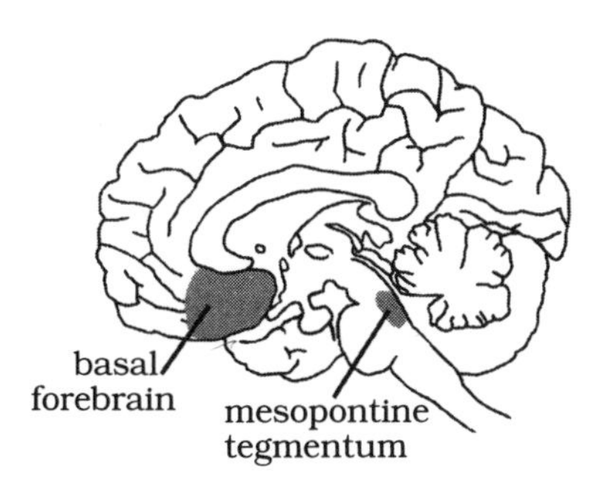

FIGURE 1.14
Acetylcholine sites

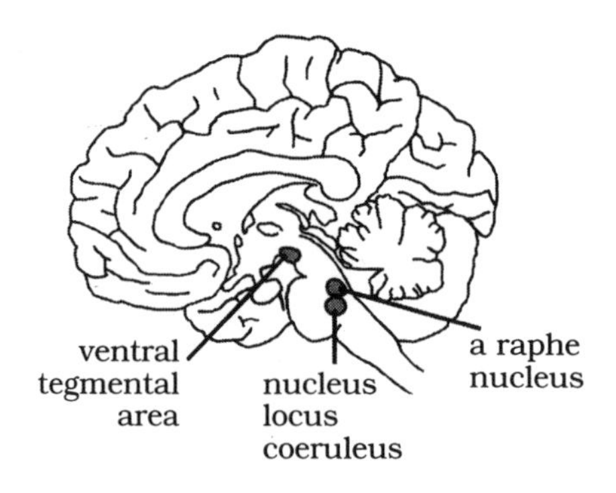

FIGURE 1.15
Serotonin, norepinephrine, and dopamine sites

Figure 1.15). These neurons produce **serotonin** (5HT) and deliver it widely in the forebrain. Serotonin is well known for its use in antidepressant medications—the SSRIs (*selective serotonin reuptake inhibitors*). This phrase allows us to expand slightly our knowledge of cellular neurophysiology discussed earlier. Recall that a neurotransmitter is excreted via an axon into the synaptic space, where it attaches to receptors on the next neuron and thereby increases or reduces its firing rate. An additional fact is that the neurotransmitter is not lost in the second cell. After a period of time, the neurotransmitter is absorbed back into the first cell, so that it can be reused. This process of retrieving the neurotransmitter is called "reuptake." SSRIs are reuptake *inhibitors*, which implies that they slow the process of reabsorbtion of the neurotransmitter back into the first cell. This means that the excreted neurotransmitter is active in the synaptic space for a longer period of time and excites the second neuron accordingly. Any chemical that inhibits reuptake of a neurotransmitter has the effect of rendering the neurotransmitter (in this case, serotonin) more effective by making it longer-lasting.

The third class of neurotransmitter that has its origin in a core brainstem nucleus is called **norepinephrine** (NE; known as

noradrenaline in Britain). This neurotransmitter has its source cells in the **nucleus locus coeruleus** of the pons (Figure 1.15). As with other state-dependent systems, the sites of action of this system are extremely diverse. (This system is discussed again, in chapter 6, in relation to sleep and dreaming.)

The last of the neurotransmitters to be mentioned here is produced in a transitional region between the midbrain and the diencephalon called the **ventral tegmental area** (Figure 1.15). The neurotransmitter produced by these cells is called **dopamine** (DA). Dopamine is also produced in other sites in the brainstem, the best-known of which is the **substantia nigra** (well known due to its role in Parkinson's disease). This nucleus is the source of the nigrostriatal DA system (which projects mainly onto the basal ganglia), but the system that originates in the ventral tegmental area is more important for our purposes. This is called the *mesocortical–mesolimbic* DA system, because it acts principally on limbic and cortical structures on the medial surfaces of the forebrain. Its main targets are the *hypothalamus*, **nucleus accumbens** (a basal-forebrain nucleus nestled beneath the basal ganglia), *anterior cingulate gyrus*, and *amygdala*. This system also projects to other structures, including the frontal lobes as a whole.

The neurotransmitters just described (as well as others that we have not discussed, such as **histamine**, which is sourced mainly in the hypothalamus) are called **neuromodulators**. This refers to the fact that the state-dependent neurotransmitter systems through which they operate exert global effects, via mass-action, over and above the existing activities of the specific pathways in the channel-dependent systems. They *modulate* these activities, in response to the current state of the organism. Thus, for example, all cognitive operations are affected—in a relatively global way—by changes in mood, vigilance, and waking state.

Neuromodulators that are not neurotransmitters

Thus far, communication between cells has been described exclusively in terms of the mode of action of classical neurotransmitters. However, there are other ways that neurons communicate with each other and with other cells of the body.

The first of these is more primitive in its delivery mechanisms than the elegant sophistication of neurotransmitters acting at synapses. However, this type of specificity is not required for systems that are state-dependent and therefore act globally on the brain as a whole. This applies above all to **hormones**. Hormones include some well-known chemicals like **estrogen** and **testosterone,** which are linked with the sexual changes of puberty and reproduction (see chapter 7), and **steroids**, which play an important part in *stress* reactions. These chemicals are produced in the viscera and are transmitted upward to the brain through the circulatory systems of the body (via the bloodstream and the cerebrospinal fluid, which constantly pump through the central nervous system as a whole). Hormones are also produced *by* the brain (mainly in the pituitary gland, which is governed by the hypothalamus) and released from there into the same circulatory systems, thereby influencing the viscera "from above." This is one of the principal "motor" channels of the brain regions that are directed toward the internal world.

The last class of neuromodulators are the **peptides** (or neuropeptides). As with hormones, peptides are produced not only in the brain but also in the visceral body. However, these substances differ from hormones in that they act over comparatively short distances. Like hormones, they produce many of their effects by a direct process of *diffusion*, bypassing the synapse, thereby modulating the firing rate of neurons already activated (or inhibited) by the classical neurotransmitters. Thus, like hormones, peptides are relatively blunt instruments. However, we should point out that while the delivery system of these neuromodulators is comparatively primitive, the chemical consequences of their operation can be highly exact, and in this respect it is a misnomer to call them primitive or nonspecific. They do not act on the brain in a *functionally* unselective way. Neuropeptides have highly specific modes of action, often closely linked to single emotional systems (see chapter 4). In fact, they are far more specific in this respect than the neurotransmitters discussed above, which are the focus of current psychopharmacology. The prospect of manipulating the emotion-specific neuropeptide systems holds great promise for psychopharmacology in the future. Varieties of neuropeptides come in the dozens, indeed the hundreds, and more are being discovered all the time. Some of them are introduced by name in later chapters.

A METHODOLOGICAL COMMENT

At the end of this introductory discussion about brain anatomy, physiology, and chemistry, a brief comment on the status of evidence in neuroscience is appropriate. The factual material in this chapter, and indeed in later chapters, represents very concrete and practical knowledge. For someone with a background in the humanities, it may be of interest to consider the extent to which such facts are "known"—how permanent are these findings, and to what extent do they still remain open to interpretation? By and large, the body of knowledge just summarized is based on extremely strong evidence. This is the beauty of neuroscience, and this is why it has so much to offer to psychoanalysis and related disciplines. As a reasonable generalization (and hoping that psychotherapists will not take offense), knowledge in psychoanalysis is highly theoretical, as there is a great deal of ambiguity between finding and hypothesis, between observation and interpretation, and, indeed, between discovery and invention. In the neurosciences these dividing lines are much clearer.

This is primarily because we are dealing with tangible, physical *things* in neuroscience. One can (usually) literally "see" what is being talked about: "This is GABA, this is glutamate, hence this stuff is going to excite, and this stuff is going to inhibit this cell; okay, let's see if it does . . . yes, as predicted, the neuron was excited by the glutamate."

With such transparent conceptual and technical tools at our disposal, we can discover facts in a secure way. This is not to say that neuroscience does not have its controversial aspects. Certainly it does. However, the bedrock of its knowledge is generally agreed upon. Moreover, where there is controversy, neuroscientists can devise and execute critical *experiments* to test who is right and who is wrong. Typically (after some debate about whether the experiment was the correct test or not!), the losing side agrees that they were wrong. Thus, the field grows, building on increasing areas of universally agreed on (and replicable) findings. All of this is due to the fact that the evidence that underpins theory in neuroscience is relatively unambiguous. The situation in psychoanalysis is different. As a general principle, psychotherapists deal with subjective experiences in the raw—that is to say, they deal with real-life stories, made out of feelings, thoughts, and memories, as they unfold in the complexities of a relationship. Subjective experience is a fleeting and fugitive thing, which makes experimental verification extremely difficult (though not, in principle, impossible). Because key experiments are so difficult to devise in psychoanalysis, and the evidence is so seldom clear-cut, the field has tended to fractionate into groups based around strongly held theoretical positions. In most cases, more than one group cannot be correct on a given point, but no one appears to be able to provide a critical test that is sufficiently robust to convince the other groups that they are wrong.

Thus neuroscience, by virtue of the objective status of its evidence, provides a useful set of anchor points from which to reevaluate psychoanalytic concepts. And psychoanalysis, by virtue of its rich theoretical tradition, provides neuroscientists with a comprehensive conceptual framework to guide their research into the neuroscience of subjective experience. This will not transform subjective experience into physical substance. There are some things that can never be seen with our eyes, though this does not make them any less real. However, *linking* the invisible inner world of feelings, thoughts, and memories with the visible tissues of the body that generate them renders them far more accessible to scientific scrutiny. It also deepens immeasurably the value of what we *can* see with our scientific eyes.

The next chapter takes us a little further in our consideration of the basic question of how the anatomical and physiological material discussed so far can actually "produce" a sentient being.

CHAPTER 2

MIND AND BRAIN—HOW DO THEY RELATE?

Although the subject matter of this chapter is still introductory, it will almost certainly be of more interest to our readers than the basics of brain anatomy and physiology. In this chapter, we consider the relationship between mind and brain in general.

One of the main points to emerge from the previous chapter was that the brain is simply a bodily organ, like the stomach, the liver, or the lungs. It is tissue, made of cells. These cells do have some special properties, but they are by no means magical. Nerve cells are of roughly the same type and employ roughly the same sort of metabolic and other processes as other cells in the body. And yet the brain has a special, mysterious property that distinguishes it from all other organs. It is the seat of the mind, somehow producing our feeling of *being* ourselves in the world *right now*. Trying to understand how this happens—how matter becomes mind—is the **mind–body problem**.

The mind–body problem is a philosophical conundrum that dates back to classical antiquity, and probably beyond. What has changed in recent years is the emergence of a comprehensive *scientific* effort to solve this ancient problem. This effort, which involves neuroscientists, psychologists, and even philosophers, takes the form of a multidisciplinary enterprise called

cognitive science.[1] In different ways, all are trying to solve the same great mystery. The advent of science to the problem has changed it slightly, in that the mind–body problem is now commonly described as the problem of "**consciousness**." In other words, the problem, "*how does the mind emerge from the brain,*" has become, "*how does consciousness emerge from the brain.*" Although psychoanalytically minded readers need no reminding that mental life is not synonymous with consciousness, we will not address this particular twist to the problem just yet. For now, let us assume that the two ways of putting the problem are synonymous.

Investigating consciousness has become the second career of Francis Crick, the Nobel Prize-winning biologist famous for being the codiscoverer (in the 1950s) of the double-helix structure of DNA. In a recent book, entitled *The Astonishing Hypothesis*, he puts into the following words what was proposed in the second paragraph of this chapter:

> The astonishing hypothesis is that you, your joys and your sorrows, your memories and your ambitions, your sense of personal identity and free will are, in fact, no more than the behaviour of a vast assembly of nerve cells and their associated molecules. [Crick, 1994, p. 3]

This hypothesis seems self-evidently true, and yet it is something that many people do not find easy to accept. How can all this—all that comprises *you*—be reduced to the activity of a group of cells? The subtitle of Crick's book is *The Scientific Search for the Soul*. This (perhaps overstated) phrase captures something of the magnitude of the problem. The individual cells

---

[1] *Cognitive* science is, in our opinion, an unfortunate term in that it implies an exclusion of noncognitive mental functions such as emotion and motivation. Nevertheless, we employ the conventional term in this book because it is so widely used. For further discussion see Turnbull (2001).

of the brain are not uniquely "mental," yet when they are connected up together, each one contributes something to something else that somehow becomes the mind.

THE "EASY" AND THE "HARD" PROBLEM

David Chalmers—one of the philosophers participating in the interdisciplinary field of "cognitive science"—argues that one aspect of the mind–body problem is "easy" and the other "hard." (Chalmers, 1995). In this way, he divides the issue into two separate problems.

The **easy problem** is the one that most neuroscientists are concerned with, and it is the one discussed by Crick in his *Scientific Search for the Soul.* Crick attempts to solve the problem by neuroscientific means. His research strategy is to try to find the specific neural processes that are the *correlates* of our conscious awareness (he calls them the "neural correlates of consciousness," or *NCC* for short). Finding the neural correlates of consciousness is a problem of the same general type as finding the neural correlates of anything—language or memory, for instance. Neuroscience has made great progress in solving such problems in the past. Finding the brain regions and processes that *correlate with* consciousness is simply a matter of directing an existing research strategy from areas of previous success (language, memory) onto a different aspect of mental functioning (consciousness).

We should not underestimate the difficulty of finding the neural correlates of consciousness, but Crick is only looking for *which* brain regions or processes correlate with consciousness and describing *where* they reside. He does not attempt to explain *how* that particular pattern of physiological events makes us conscious. This is the **hard problem**. The hard problem is a conundrum of a different magnitude—it raises the question of how consciousness ("you, your joys and your sorrows, your memories and your ambitions, . . .") actually emerges from matter. Modern neuroscience is well equipped to solve the easy problem, but it is less clear whether it is capable of solving the hard problem. Science has few precedents for solving a problem that philosophers have deemed insoluble *in principle*.

John Searle, another contemporary philosopher with a great interest in this problem, suggests the following thought experiment (1995a, p. 62).[2] Pinch yourself (hard) on your left hand. What happens? You feel pain, of course—it is sore. This is an expression of the mind–body problem: something *physical* happened to your hand, and yet you *felt* a pain in your mind. Let us see, in terms of the *easy* problem, how we fare in explaining this phenomenon.

We know exactly what the pain receptors embedded in your skin look like, and how they work. When pressure is applied to these receptors, a very specific physical process excites the neurons connected with them (Figure 2.1). This sends a message down those neurons (causes them to fire), which in turn causes a chemical to cross the synaptic spaces at the ends of the axons—using the channel-dependent neurotransmitter systems discussed in chapter 1. The axons in question travel through a nerve coursing up the arm into the spinal cord, and then upward through the various parts of the spinal cord and the brainstem (seen in section in Figure 2.1) along its dorsal column. Crossing the midline in the brainstem, these axons terminate on a second set of neurons in the thalamus. From there the physiological message is relayed again, to a specific part of the primary sensory cortex of the right hemisphere (again using the method of transmission described in chapter 1). The pain receptors in the

[2] A "thought experiment" is an imaginary experiment; the experiment is not really conducted. Searle (1995a, 1995b) provides a highly readable overview of the general problem under discussion in this section.

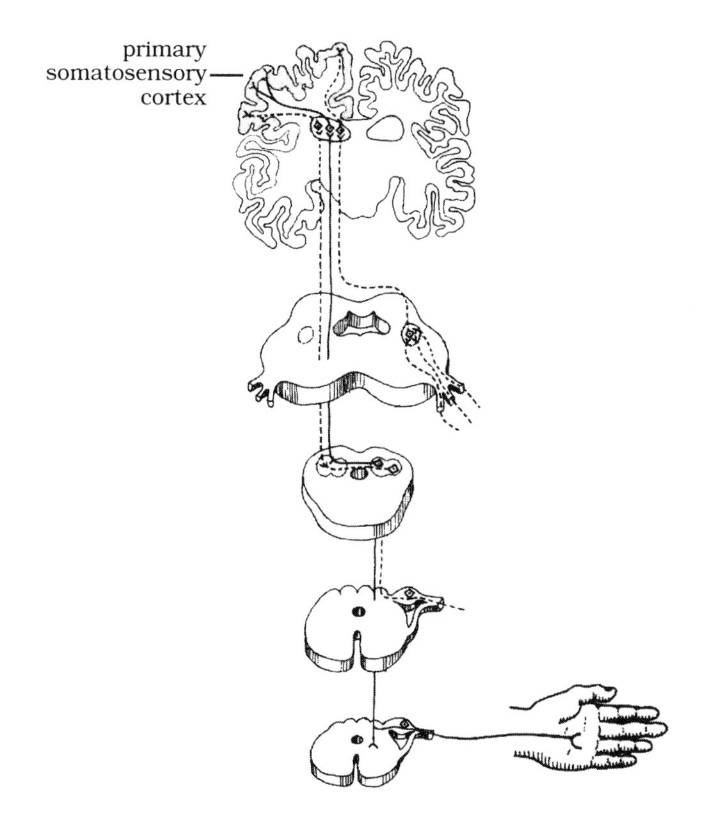

FIGURE 2.1
Schematic representation of somatosensory pathways

left hand are represented in a specific region of the somatosensory cortex in the parietal lobe, and that is where the nerve fibers we have been tracing terminate. (Pain receptors from other parts of the body map to different regions in the somatosensory cortex, as suggested by the dashed lines.) Excitation of the cortical cells in this area causes you to feel pain. This solves (this particular instance of) the easy problem—*these* are the physiological processes that cause you to feel pain in your hand.

But it is not difficult to see that the hard problem remains entirely unsolved. What turned the physiology, anatomy, and chemistry just described into a feeling of pain? How did *that* happen? We have just outlined a purely physiological process (and traced the anatomical pathways it traversed); we have not explained how the process started as something physical but somehow ended up as something mental. Searle (1995a, p. 62) used a memorable phrase to describe the hard problem we are left with: "How does the brain get over the hump from electrochemistry to feeling?"

This sort of question was traditionally considered to be a philosophical problem, but it is now being treated as a scientific one—one that might be addressed experimentally. To approach this transformation of the problem, and what it reveals, we begin here with a brief overview of the classical *philosophical* approaches to it, followed by an equally brief history of the approach taken by *neuropsychology*—the scientific discipline specifically devoted to investigating relationships between the mind and the brain. We shall see that while philosophers relish Chalmers' hard problem, neuroscientists have for the most part not yet really tackled it.

MATERIALISM AND IDEALISM

Perhaps the most basic distinction among philosophical approaches to the mind–body problem is the one that divides materialists and idealists. The **materialist** position, represented pictorially in Figure 2.2, is that *everything* is ultimately reducible to matter. From this standpoint, the thought on the right of the diagram does not really exist. Its existence is *illusory;* the mind is *really* an aspect (or function) of matter (the left side of the diagram).

At the other extreme, the **idealist** contends that only the mind really exists (for us, at least). For all the apparent substantiality of matter, the "things" we see, touch, and hear are *really* nothing other than products of our mental processes (i.e., they

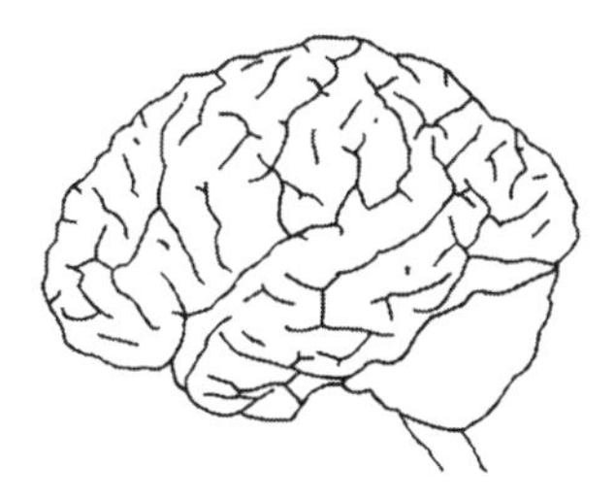

FIGURE 2.2
Brain and mind

are actually *perceptual images*). We can never reach beyond the envelope of conscious awareness and demonstrate the existence of any thing independent of our mind's perceptual images. So, from this standpoint, the concrete thing on the *left* of the diagram in Figure 2.2 does not really exist, or at least it, too, belongs inside the "thought bubble" on the right.

Although these positions both appear logically tenable, the idealist position has fallen out of favor. This seems to be due to the vagaries of intellectual fashion more than anything else. Within cognitive science today, just about everyone is a materialist. But materialists come in different shapes and sizes.

MONISM AND DUALISM

The dichotomy between monism and dualism is perhaps just as fundamental as the one between materialism and idealism, and is easily confused with it. According to the **monist** position, we are made of only *one* kind of "stuff." In other words, mind and matter (which appear to be two things) are really reducible to one and the same thing. This might seem to be identical to the materialist position just described (and the two arguments do normally go together), but the monist position does not actually state that the singular stuff we are made of is *matter*. A monist could just as well claim that we really consist only of *mind* (thereby embracing an idealist position) or even that we are actually made of some *other* kind of stuff, as yet undefined, which is neither mind nor matter. In the monist position, all that counts is that the apparent distinction between mind and matter dissolves into a common something.

The **dualist** view—closely associated with the name of René Descartes—simply states the opposite: We are divided in our essence and are made of *two* kinds of stuff. Matter and mind (or body and soul) are quite irreducible to one another. Like idealism, dualism is very unfashionable nowadays. Most cognitive scientists, therefore, are *materialist monists*: they believe that mind and brain are ultimately reducible to a *single* kind of stuff, and that that stuff is *physical*—specifically, some property of neurons (or an aggregate or subset of neurons).

REDUCTIONISM, INTERACTIONISM, AND OTHER STRANGE THINGS

Materialist monism defines the *relationship* between two types of stuff. On this view, one type of stuff (brain tissue) is more fundamental, even more real, than the other (conscious awareness). In Crick's statement—"you, your joys and your sorrows, your memories and your ambitions, your sense of personal identity and free will are, *in fact, no more than* the behavior of a vast assembly of nerve cells"—"you" are *reduced* to nerve cells. The essence of Crick's **reductionism** resides in the words we emphasized: "in fact, no more than." Reductionism *reduces* one thing to another (in this case, mind to brain) and thereby *explains it away*. However, not all materialists are reductive (see below).

Dualists are, by definition, nonreductive. The crux of their position is that mind and brain *cannot* be reduced to one another. What, then, is the nature of the relationship between the two? A dualist's answer to this question determines what *kind* of dualist he or she is. Most dualists describe the relationship between mind and body in **interactionist** terms; they assert that physical events have mental effects, and vice versa. The interactionist view, then, is simply that body and mind *interact* with each other. This seems perfectly plausible and appears easy to demonstrate empirically: plummeting blood sugar causes loss of consciousness (physical event causes mental event); freely deciding to move your hand causes it to move (mental event causes physical event). But when the logical underpinnings of this dualist position are spelled out, it seems less plausible: the interactionist actually claims that bodily *stuff* and mental *stuff* interact with each other. This way of putting it immediately reveals the pitfalls of almost any dualist position. How, exactly, does a thought (which has no physical properties whatever) cause the physical stuff of neurons to start firing? This violates all the known laws of physics.

Other varieties of dualism fare no better. One such well-known variety is called psychophysical **parallelism**. This position avoids some problems of interactionism by claiming that mental and physical events do not have a *causal* relationship; the two classes of event simply *co-occur*—they are *correlated* with one another. Whenever something specific happens in the brain, something equally specific happens in the mind, and vice versa. The two things occur together, in unison. If the basis of this correlation still seems mysterious, that's because it is. The parallelist does not feel obliged to *explain* this linkage.

EMERGENCE

We said above that not all materialists are reductionists. Many cognitive scientists today hold the view that the mind is an **emergent property** of the brain. According to this view, mind and brain are equally *real*, but they exist at different levels of complexity. Just as water (wet and flowing, at room temperature) emerges from a particular combination of hydrogen and oxygen and has distinctive properties of its own (properties that do not characterize hydrogen or oxygen alone), so, too, mental phenomena emerge when the neurons of the human brain are connected or activated in a particular way. The mind can thus be regarded as a higher level of organization of neurons, just as water is a higher level of organization of its constituent atoms. The problem with this apparently sensible argument is that it does not really explain the *mind–body* relationship; it simply equates the relationship between mind and body with another type of relationship in which the problem at hand does not exist. It is no great achievement to explain how *matter* "gets over the hump" from hydrogen and oxygen to water.

THE LIMITS OF KNOWLEDGE

It is possible to find some merit in all of these different philosophical positions. It is also possible, with a little effort, to make all of them look ridiculous. This might be one good reason to replace the philosophical approach to the "hard" problem with a scientific one. As we said at the end of chapter 1, competing positions are pitted against each other in science and *tested* experimentally to determine which one is correct. *But not all propositions are testable.* For example, how can one test the proposition, "God exists"? Loath as we may be to admit it, the testable hypotheses that scientists can work with are embedded in sets of broader propositions that are themselves untestable. These broader propositions define the world-view (*Weltanschauung*) within which a scientist operates; and world-views are not subject to proof. Science is limited to answering questions that can be asked *within* a particular world view; it cannot test the world view itself.

It has yet to be determined whether the various philosophical positions on the mind–body problem constitute "world-views" in this sense, or whether they will someday (perhaps soon) be transformed into testable hypotheses. We [MS, OT] are of the opinion that the nature of the relationship between brain and mind (body and soul) is *not* amenable to scientific proof. Statements such as "body and soul are one" (the monist position) or "the soul does not really exist" (the materialist position) are not, in our view, scientifically testable statements. They are of the same order as the statement, "God exists." We believe that scientists can do no more than ensure that they are *aware* of the world-views they endorse, because the assumptions these entail will determine the experimental questions they ask and how they interpret them.

We have said already that most neuroscientists working on the mind–body problem today (i.e., on the problem of "consciousness") endorse a materialist–monist position. In other words—whether they recognize it or not—they *assume* that mental life is the product of a vast assembly of neurons; then they set about determining which processes in that assembly "cause" consciousness. Note the problematic status of the word "cause" in this context. This graphically illustrates how important it is for scientists to be aware of the philosophical positions they have adopted. It is appropriate to describe certain neuronal processes as *causing* consciousness *only within a particular philosophical framework*. Even if a particular subset of neuronal processes is experimentally proven to be uniquely associated with conscious experience, it still remains possible (within a dualist framework, for example) to view this association as a *correlative* rather than a causal one. For this reason—because the assumption that neuronal processes "cause" consciousness *begs the very question* that the "hard problem" poses for science—we do not endorse the materialist position that most of our colleagues currently adopt. We favor a slightly different position, one that seems more agnostic and leaves the possibilities more open.

DUAL-ASPECT MONISM

Dual-aspect monism accepts that we are made of only one type of stuff (that is why it is a *monist* position), but it also suggests that this stuff is *perceived* in two different ways (hence, *dual-aspect* monism). The important point to grasp about this otherwise straightforward position is that it implies that *in our essence* we are *neither* mental nor physical beings—at least not in the sense that we normally employ these terms. This requires some explaining. Dual-aspect monism (as we understand it) implies that the brain is made of stuff that *appears* "physical" when viewed from the outside (as an object) and "mental" when viewed from the inside (as a subject). When I perceive myself externally (in the mirror, for example) and internally (through introspection), I am perceiving the *same thing* in two different ways (as a *body* and a *mind*, respectively). This distinction between body and mind is therefore *an artifact of perception.* My external perceptual apparatus sees me (my body) as a physical entity, and my internal perceptual apparatus feels me (my self) as a mental entity. These two things are one and the same thing—there *really* is only one "me"—but since I am the very thing that I am observing, I perceive myself from two different viewpoints simultaneously. This problem does not arise when we observe other things, since those other things are not our*selves*.

What, then, are we *actually* made of? This is the big question that dual-aspect monists ask of science. We can never literally *perceive* the stuff we are made of without first representing it through one of our perceptual modalities—which means that we can never escape the artificial mind–body dichotomy. Since we can never transcend the limits of our senses, we can never perceive the *underlying* mind–body stuff *directly*. We can only make *inferences* from the data of perception (from scientific observation) as to the nature of that underlying entity—let's call it the "*human mental apparatus*"—and inferences about how it is constructed and how it works. Our picture of the mental apparatus *itself* will therefore always be a figurative one—a model.[3] We possess concrete perceptual images of its two observable manifestations (the brain and subjective awareness), but the underlying entity that *lies behind* those perceptual images will never be directly observable. Scientific observation has its limits.

This is not a unique situation. There are many things that scientists are concerned with that cannot be perceived directly. Witness, for example, the "quarks" of contemporary physics, or even the force of "gravity." Nobody doubts the existence of these ultimate things, yet they can only be observed via their perceived *effects*. What makes the mind–body problem unique is only the fact mentioned a moment ago: when it comes to the human mental apparatus, the observer *is* the instrument that he or she uses to observe it. If we accept that the mind–body problem thereby boils down to *a problem of observational viewpoint*, and that the distinction between your self and your body (between mind and matter) is therefore merely an artifact of perception, the "hard problem" evaporates. Then we are left only with the "easy" problem—namely, which brain processes *correlate with*

[3] Freud described this type of model building as "metapsychology." This term refers to our attempts to see beyond [*meta*] consciousness [*psyche*]. Freud contrasted metapsychology with *metaphysics*, which is a branch of philosophy concerned with similar problems, but which attempts to solve them through pure reason rather than scientific observation and experimentation.

which subjective processes. We can then also ask: What can be inferred from these two (correlated) sets of data about the functional organization of the underlying apparatus that generates them? In this context, where the laws governing the apparatus itself have to be *inferred* from the observable data, combining the knowledge derived from both observational perspectives is far better than relying on either perspective alone.

Keeping all these philosophical points in mind, then, let us move on and consider what the history of neuropsychology teaches us about *scientific* approaches to the mind–body problem.

WHY THE BRAIN?

In antiquity, the heart and other organs like the stomach were believed to be the seat of the soul.[4] This may be because we experience visceral sensations, such as increased heart rate or "butterflies" in the stomach, in relation to some emotions. We have no direct record of how the emphasis shifted to the brain, but we can hazard a pretty good guess. The comments of Hippocrates and other classical scholars suggest that they may have settled on the brain because of *clinical observations*. When people suffered internal damage to the head—such as through a direct blow to the skull in warfare—their minds were altered, just as happened in the case of Phineas Gage (chapter 1). This occurred often enough, and was witnessed by enough medical observers in the ancient world, for them to realize that there is something special about the brain that relates it to the mind in a way that does not occur with any other organ.

[4]These beliefs still survive in everyday figures of speech, such as "I have a gut feeling," "I believe it with all my heart," "she broke my heart," and so on.

THE CLINICO-ANATOMICAL METHOD AND "NARROW LOCALIZATION"

Observations of the type just referred to were gradually formalized into a basic tool of scientific medicine, known as the **clinico-anatomical method**. This method was explicitly introduced into neuroscience about 150 years ago, in Paris, by Jean-Martin Charcot, the world's first professor of neurology. The method involves systematic correlation of mental functions that are changed (clinically) with (anatomical) damage to particular areas of the brain. The goal is to establish lawful (clinico-anatomical) correlations between the different mental functions and the different parts of the brain. Such correlations, as we shall see, teach us a great deal about the underlying organization of the mental apparatus.

Pierre Paul Broca, a French anthropologist and physician, is credited with the first real breakthrough in this regard. He looked after a patient called Eugène Leborgne, who had lost the power of speech, in a Paris hospital. Before acquiring his neurological disorder Leborgne had been healthy, but his language abilities declined after the onset of his disease, until he was no longer able to utter meaningful words or phrases. Today, we call this disorder **aphasia**.

The only thing Leborgne could say was "tan," which he uttered *ad nauseam*, and because of this he became known to the hospital staff as "Tan-Tan." After Leborgne died, an autopsy revealed a brain lesion that affected mainly the left inferior (i.e. lower) frontal lobe. As a result of this case, and a few others with the same symptoms associated with damage in the same place that Broca reported four years later, he was able to announce in 1865 that he had found the brain "center" for language—a specifically human mental function. Other investigators later confirmed that this area is located in the posterior and inferior

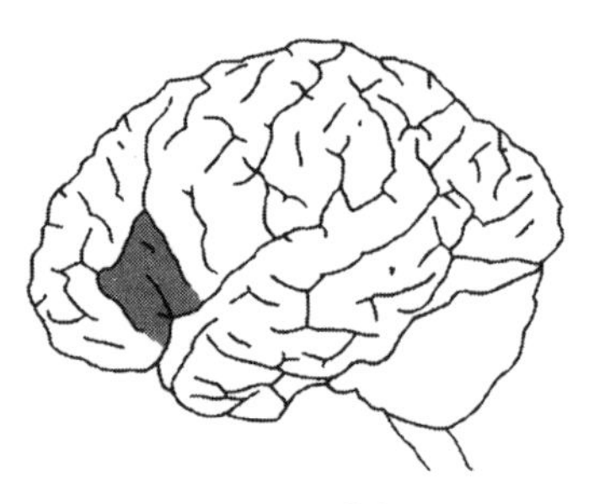

FIGURE 2.3
Broca's area

region of the left frontal lobe (Figure 2.3). This part of the brain has been known as **Broca's area** ever since.

Following Broca's discovery, other European neurologists made a sequence of such clinico-anatomical correlations, in connection with other mental functions. One found a center for recognizing objects, another for calculation, another for skilled movements, and so on. It soon became clear that these functions could be divided into component parts. Thus, for example, it was learned that Broca's area was not the center for language *as a whole*, but rather for the ability to *produce* language; another area (known as **Wernicke's area**) is specialized for the *comprehension* of speech, yet another for the *repetition* of speech, and so on (see chapter 9). On this basis, a map (or mosaic) of specialized regions of the cortex was constructed over the course of a few decades. The quest to locate the different brain regions associated with particular mental functions became known as **localizationism**.

OPPOSITION TO LOCALIZATIONISM

The localizationist era was a golden age in the history of neuroscience. The movement had its opponents, however. These opponents believed that linking psychological functions with specific brain areas created a distorted picture of mind–brain relations, and they derided the localizationist maps as "brain mythology." The extreme alternative approach was **equipotentialism**. The equipotentialist school argued that it is not *where* the brain is damaged that counts, but rather *how much* brain is damaged. They argued that the more of the brain is damaged, the more of the mind is lost.

Some of the disagreement between these two schools appears to have derived from the fact that they studied different populations. The equipotentialist position was derived, in large part, from experiments on *birds*. These experiments showed that the more brain was damaged, the poorer the bird's performance was on almost any task. The problem with investigating bird brains is that they are not a good source of information on higher mental functions. For one thing, the brains of birds are very small and therefore not ideal for making precise anatomical distinctions (especially using the surgical techniques of the time). Nevertheless, the equipotentialist school raised some important questions for the localizationist approach. For example, it is true that larger lesions in humans cause psychological deficits that are more than a sum of the deficits produced by a number of corresponding smaller lesions. Although the classical case reports often described patients whose lesions were very large, and therefore affected many functions, the investigators of those cases would focus on only *one* of the patient's psychological deficits. In the case of Broca's patient, Leborgne, the lesion involved a huge region of the brain—by some accounts as much as *one-third* of the left hemisphere.[5] Broca chose to focus on the

---

[5] Broca actually argued that the damage in his case was *most severe* in what is now called Broca's area, suggesting that a degenerative process had *begun* there. Those who had known Tan-Tan for years after his accident said that his first problems had been in language. From this, Broca concluded that the most badly damaged part of the much larger lesion was responsible for Tan-Tan's language disability. In retrospect, we know that Broca was basically correct in this judgment. However, his original, celebrated localization appears to have been mostly (inspired) guesswork.

language deficit, but this probably would have been just one of a catalog of psychological disorders from which this poor man suffered.

In summary, the early history of neuropsychology was characterized by two polarized views. The localizationist view was that the mental apparatus was constructed of a network of centers connected together in a mosaic-like pattern. Each of these centers was the neural correlate of one component of the mind. When connected together, you would have the whole mind. The opposing view was that psychological functions (such as language, memory, and arithmetical ability) were the result of the entire brain working together. It was therefore not possible to attribute whole mental functions to circumscribed parts of the brain. The brain, like the mind, formed a functional unity.

SOME INTERESTING LINKS TO PSYCHOANALYSIS

Gradually, as the discipline of neuropsychology progressed, it settled on a *third* view. The new view was an amalgamation of the original positions. One of the first people to promulgate the beginnings of this third view was an obscure Viennese neuroscientist named Sigmund Freud. Freud was a neuroanatomist turned clinical neurologist with an interest in aphasia, cerebral palsy, and the psychopharmacological properties of cocaine. In 1891, Freud published a book entitled *Zur Auffassung der Aphasien* [Conceptualizing the Aphasias],[6] in which he put forward a brilliant critique of the localizationist theory of language while keeping his distance from the extreme equipotentialist alternative. The locationalist doctrine was in full flourish at that

[6]This book was translated into English in 1953 under the title *On Aphasia* (Freud, 1891b). A new translation will appear soon (with a revised title) as part of the four-volume *Complete Neuroscientific Works of Sigmund Freud.*

time, however, and few neuroscientists took much notice of Freud's book (although today it is regarded as a classic).[7]

Freud's arguments only gained credence when they were reintroduced 70 years later. This occurred in 1966 with the publication of a book called *Higher Cortical Functions in Man*, written by a Russian neurologist named Aleksandr Romanovich Luria (who divided the brain into the three "units" that we mentioned—in passing—in chapter 1). Luria in fact had close connections with psychoanalysis in the 1920s (see Kaplan-Solms & Solms, 2000; Solms, 2000b). In *Higher Cortical Functions* (and some earlier books, which were less influential), Luria introduced the concept of **functional systems** to neuropsychology. It was this concept that finally reconciled the localizationist and equipotentialist positions. Luria had an enormous influence on modern neuropsychology, and he still remains one of the most highly cited authors in the discipline today (for a review see Turnbull, 1996).

WHAT ARE FUNCTIONAL SYSTEMS?

Luria agreed with the equipotentialist view that it is incorrect to assert that centers such as Broca's area "produce" mental functions like language. However, he also agreed with the localizationists that different brain regions do have special (indeed, unique) functions. He concluded that the disagreement between the two traditional viewpoints arose from an overly narrow definition of "*function*." To clarify, Luria pointed out that many bodily functions are the products, not of one particular tissue, but, rather, of an interaction *between* a number of different

[7]According to Ernest Jones (1956, p. 237), 850 copies of this 1891 book were printed. Only 257 had been sold after nine years, so the remaining books were pulped.

tissues. For example, *digestion* is not a function of the stomach alone. It is misleading to say that digestion is "produced" by the stomach, just as it would be misleading to say that it is produced by the liver, pancreas, and bowel (to mention just a few of the other organs involved in digestion). All these structures *together* perform the complex function of digestion. *This* is a functional system. The same principle applies to other complex functions. *Respiration*, for example, is not "produced" by lung tissue; it arises from an interaction between the lungs, the intercostal musculature, cardiovascular circulatory processes, and nervous-control mechanisms, among other things. Respiration is thus the product of a complex functional system.

Luria argued that the neurological organization of mental functions is no less complex than are digestion and respiration; accordingly, there are no neuroanatomical "centers" for the psychological functions of the mind. Mental functions, too, are the products of complex systems, the component parts of which may be distributed throughout the structures of the brain. The task of neuroscience is therefore not to localize "centers," but, rather, to identify the *components* of the various complex systems that interact to generate the mental functions. Luria called this task "dynamic localization." Readers will learn in subsequent chapters of this book that functions such as emotion (chapter 4), memory (chapter 5), and dreaming (chapter 6)—and indeed consciousness itself (chapter 3)—are all generated by distributed networks of brain structures such as this, with each structure contributing a particular component to the dynamics of the system as a whole. Disturbance in any of the component parts causes the entire functional system to malfunction, but in different ways, depending on which part of the system is damaged. These are the symptoms that neuropsychologists study. Mental functions, in short, are not localized *in* any of the component structures, but rather *between* them. Like the mental apparatus as a whole, they are *virtual* entities.

FUNCTIONAL SYSTEMS AND THE EASY PROBLEM

By the 1970s, neuropsychology had settled on the idea that the neural correlates of mental functions are complicated functional systems, and this is more or less where we still stand today. One difference is that we have a more (and increasingly) detailed understanding of the operations contributed by the different component parts. Thus, for example, we now know not only that the hypothalamus is an important component of the functional system for emotion; we also know much more about how the hypothalamus *itself* works. This is due to the supplementation of dynamic localization with new, high-technology techniques. The strength of the functional systems approach, based on the clinico-anatomical method, is that it allows us to "carve mental functions at their joints." Once we know which parts of the brain constitute the basic components of a particular functional system, we know a great deal about the *internal organization* of the mental function in question that we did not know before; we know, in a sense, *how it is put together*. For example, we know not only what brain regions contribute to emotional life, but also what the neural constituents of emotion in general are, how many basic emotions there are, and also what the different chemical processes are that characterize each different emotion (see chapter 4).

This is powerful knowledge. But it is still only knowledge at the level of the "easy problem." When we said a moment ago that we know "what the different chemical processes are that characterize each different emotion," we did not mean that the chemicals literally *generate* the emotions. It is more correct (but perhaps too cumbersome) to say that neuroscientists have discovered the specific neurochemical processes that *correlate with* the subjective experiences of specific emotional states. Finding the chemical *correlates* of the different emotions does not solve the hard problem, although it does reveal something important about the actual structure of the mind. But correlating subjective states with brain anatomy and physiology is not the only way that contemporary cognitive scientists elucidate the functional organization of the mind.

THE GHOST IN THE MACHINE: ARTIFICIAL INTELLIGENCE

Interest in **artificial intelligence** (AI) has grown enormously since the 1960s. This subdiscipline within cognitive science involves building (or studying) computers that mimic the functions of the brain. For some, the aim of doing this is to learn more about how the human mind works; for others, it is literally to *build* a mind. The logic of the AI approach becomes clear when viewed in the context of Crick's "astonishing hypothesis." If—as Crick so eloquently proposed—a complex interaction of neurons is all that is needed to produce human intelligence, and if the function of neurons is simply to transfer information, then the human mind is merely an information-processing machine. Computers are information-processing machines, too; therefore, a computer, too, can generate a mind. All we need to do to construct a mind is to design the hardware correctly and program it appropriately.

This AI argument is a fascinating as well as frightening proposition. It also has serious implications for the easy and hard problems. In relation to the *easy* problem, there is certainly much to be learned about the design of the functional systems of the brain by trying to *model* them on computer programs. If we can construct artificial "neural networks" (as these programs are mischievously termed) to perform humanlike intellectual functions, then we have good reason to believe that we understand pretty well how the equivalent *real* intellectual functions work. (Such knowledge, of course, has many practical applications.) In relation to the *hard* problem, the question is this: Is it really possible to *make* a mind? If it is possible, then the really big question of *how* consciousness is *actually generated* is solved. Can consciousness really emerge from an interaction of computer chips? Is it really just a matter of information processing? If so, should some machines be afforded legal and moral rights? Do they not have feelings, hopes, and dreams like ourselves?

THE TURING TEST

The **Turing test** was designed by Alan Turing, a famous British mathematician and the developer of a forerunner of the modern computer. This test provides a controlled way of determining whether or not a machine really is intelligent—that is, whether it truly has a humanlike mind. We said earlier that our mental experience arises out of introspection; we perceive the mind (as opposed to matter) by looking inwards. It is only possible to do this because we *are* our mental instrument. For this reason, it is impossible to enter the consciousness of another being; we can only *experience* our *own* minds. When it comes to other people, we have to *infer* the quality of their consciousness from external observations. This is an expression of another famous problem in philosophy, closely related to the mind–body problem. The **problem of other minds** is that we can never know *for certain* whether or not other minds really exist (not only computer minds—*any* minds, other than one's own). This problem arises for the reason already stated: We cannot enter the consciousness of another being to determine directly if it is like our own, and indeed if it exists at all. Each one of us knows for certain only that we, *ourselves*, possess consciousness. For the rest, we rely on inference. Since other people *behave* in roughly the same way as we do, and since their brains are *constructed* in roughly the same way as ours, it seems only reasonable to *infer* that they, too, possess a consciousness like ours.

The Turing test is based on this principle. In the most famous variant of the test, the examiner is asked to present a series of questions to the occupant of a closed room, who may be either a person or a computer. To avoid the (purely mechanical) problems of speech, the conversation is conducted through a keyboard and monitor. The questions typed into the keyboard are of the examiner's choosing, and the occupant of the room sends back answers to the examiner's monitor screen. There are no time limits, nor any limits to the number of questions that may be asked. If the examiner cannot reliably tell whether the occupant is human or computer—that is, cannot reliably distinguish one from the other—then the computer is given the benefit of the doubt and is granted a "mind."[8]

Many computers (i.e., computer programs) have passed limited versions of the Turing test. Does this mean that they really possess minds? Or is it simply a matter of examiners choosing their questions unwisely? Consider the computer program called "Eliza," developed in the 1970s, which acts as a psychotherapist. Eliza occasionally passes a limited version of the Turing test.

Interactions with Eliza Here is a brief conversation that one of us [OT] recently conducted with Eliza. It gives some indication of the quality of her "thinking." The computer's statements and responses are printed in italics.

How do you do? Please tell me your problem.

Why can't I be happy?

Do you think you should be able to be happy?

I think I deserve more happiness than I have at the moment.

---

[8]Wonderful examples of the Turing test in action are depicted in science-fiction films like *Blade Runner*.

Do you really think so?

Yes, life doesn't need to be so cruel.

You seem quite positive.

I suppose I am, at least some of the time.

Is that why you came to me?

I came because I need help.

Is that the real reason?

Actually I came because I heard you were a computer and not a real therapist.

Do computers worry you?

Can I ask a question?

Whether or not you can ask a question depends more on you than on me.

The question is: How do mind and brain relate?

I'm not sure I understand you fully.

It's the hard problem—how does mental life arise from physiological events?

Please go on . . .

Many "blind" examiners of Eliza are *convinced* that they are dealing with a human therapist because the quality of interaction is so good and the computer appears to show some insight. Eliza's performance seems less impressive once they are aware of the mechanism that the computer uses to achieve its results. The program was remarkably simple to devise because it was based on the "reflective" psychotherapy technique of Carl Rogers. All the program has to do is identify the key phrases in a sentence and convert them into question form, which is then reflected back to the examiner. The program also searches for key words like "because," which it will respond to with phrases that are not reflective.

The resultant interaction is different from real psychotherapeutic situations in a number of important respects. Perhaps most important is the fact that Eliza lacks the capacity to grasp the meaning or intention *behind* an examiner's statements. This raises an interesting question: What mental capacities do real psychotherapists employ to do this? Why can a computer not mimic *these* capacities? Real psychotherapists *feel* their way into their patients' hidden intentions and motives (they *empathize* with them).[9] Unless it is possible to design a computer that has *feelings*, therefore, it will probably never be possible to design an artificial psychotherapist worth its salt. And *feelings*—as we will learn in chapter 3—are synonymous with *consciousness*. This begins to suggest that it will probably never be possible to design a computer with a mind. Although most of us probably felt this intuitively all along, it is not yet completely clear *why* this should be so. So, let us proceed further.

DOES INTELLIGENCE MAKE MIND?

Our interaction with Eliza teaches us some important lessons about artificial intelligence. Firstly, it is relatively simple to produce a computer that displays some degree of *intelligent behavior* and may therefore pass the Turing test under certain circumstances. Eliza does not always pass the Turing test, but she does very well considering how incredibly simple her program is. This shows that, if the test of whether or not something has a *mind* is really reducible to a test of intelligent *behavior*, the hard problem would have been solved long ago.

But generating intelligent behavior is vastly different from generating a mind. Even though it is possible to have an inter-

[9] Many psychotherapists today use the term "countertransference" for this (empathic) function. Empathy (or countertransference) is one of the most important ways through which we come to know "other minds."

esting interaction with Eliza, few reasonable individuals would believe that her program shows evidence of *consciousness*. Eliza is mindless, in that sense of the word. The problem of the mind is therefore probably not a problem of intelligence. Many computers display intelligent behavior (they behave appropriately, even adaptively, under relevant circumstances, and thereby usefully solve many difficult problems). But a computer must be able to generate "joys and sorrows, memories and ambitions, and a sense of personal identity and free will" (to paraphrase Crick) before we are persuaded that it possesses a mind. The fact that we are not persuaded vividly illustrates the gulf that separates the "easy" and "hard" problems in cognitive science.

MIND AND CONSCIOUSNESS

Earlier in this chapter we pointed out that the classical mind–body problem has been redefined by cognitive scientists (neuroscientists, psychologists, and philosophers) as the problem of consciousness. In a recent book, *Mental Reality*, the philosopher Galen Strawson (1996) considered, from every conceivable angle, the question we are considering here: "What *is* mind?" Strawson concluded that mind is synonymous with *consciousness*. The essence of the mind for Strawson is not intelligent behavior but, rather, subjective awareness. At this point, we are ready to agree with him.

But the argument that mind and consciousness are identical is precisely the viewpoint that Freud *opposed* so strongly a hundred years ago, when he first introduced the idea of an *unconscious* mind. When Freud wrote his early psychoanalytical works, philosophers were already saying that what is essential about the mind is consciousness, and yet Freud contended that clinical observations show that consciousness is merely a (variable and superficial) *property* of the mind. He argued that the mind extends well beyond what we are conscious of, since we all display unmistakable evidence of possessing memories, intentions, and so forth, of which we are not consciously aware. Simply because we are not consciously aware of such memories, intentions, and so forth, does this mean that they are not mental? Also, though some of our unconscious thoughts may never *reach* consciousness, they do still exert an *influence* on consciousness (and purposive behavior). According to Freud, it is therefore both legitimate and necessary to include the things that lie *behind* consciousness within our conception of the mind. This conclusion seems no less compelling than Strawson's.

In fact, Freud went further. His conception of the mind was this. He believed (1940a [1938]) that just as our awareness of the outer world is derived from objects that are really external to it, and *represented* in our perception, so, too, our awareness of the things going on inside our own selves is mere perception, not to be confused with the actual (unconscious) mental processes and contents they represent. This is why we can *misperceive* our own motives, memories, attitudes, and so forth. (This is also why Freud was a dual-aspect monist; see above.[10]) For Freud, then, the mind *itself* is unconscious, and consciousness is mere perception of the mind's actual processes. This raises an obvious question: Who does the perceiving?

IS THERE A LITTLE PERSON LIVING IN YOUR HEAD?

Freud called the part of the mind that does the perceiving the "ego." (In fact, since he was German-speaking, he called it the *Ich*, which translates literally as "I" or "me"; however, his English

[10] See Solms (1997b) for a discussion of this point. Dual-aspect monism may be the only sensible philosophical position for those who accept that there is more to the mind than consciousness.

translators preferred the Latin term "ego.") Cognitive scientists love to point out that such concepts imply that consciousness is achieved by a little person—a **homunculus**—living in your head. If an explanation of consciousness has recourse to a homuncular concept, it is not really an explanation at all; it has merely shifted the problem. The question then becomes: How does the homunculus become conscious? Is there another little person inside the head of the homunculus? This logically unsatisfactory situation is called an **infinite regression**. The homunculus problem is closely related to another problem that is equally topical in contemporary cognitive science: the **binding problem**.

THE BINDING PROBLEM

We know a great deal about perceptual processes in the brain, especially visual ones. For example, we know that determining *what* a visual object is, and *where* that object is, are tasks carried out by different brain regions. (The "*what*" processing stream extends downward from the occipital lobe into the temporal lobe; the "*where*" stream extends upward from the occipital lobe into the parietal lobe; see Figure 2.4.) We also know that there are specialized systems in the visual brain for dealing with

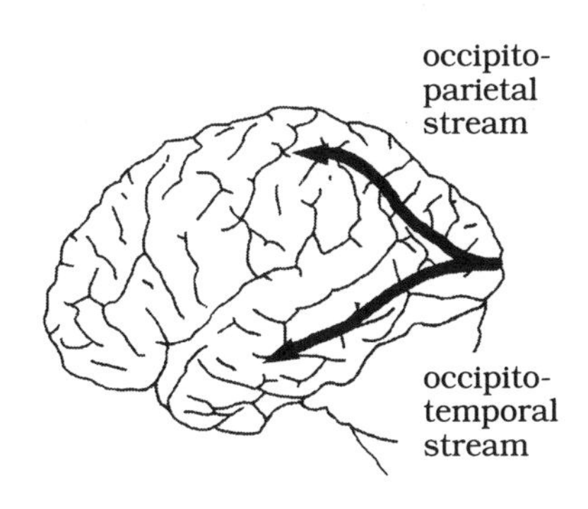

FIGURE 2.4
Two visual systems

color, with *faces*, and with *motion* information. Everything we know about visual processing—involving converging findings from neuroanatomy and neurophysiology, as well as human and animal neuropsychology—unequivocally supports the view that there is a great deal of specialization of function in human visual processing. The information arriving at the retina seems to become ever more fragmented as it progresses through the brain for further processing by these various specialized "modules." The same seems to apply to the other perceptual modalities.

These findings have some interesting implications for our understanding of consciousness. A property of consciousness that can be easily agreed upon is that it is normally a *unified* experience. Each one of us has the impression of being a single entity, experiencing an integrated perceptual world at this particular moment in time. It is *me* that is surveying the scene in front of me right now. I am the unifying point of reference for all the objects I see scattered before me. The objects, too, appear to me to exist in a unified field of space and time. That thing over there is a coffee mug. The mug appears red, has a handle, and stands on my desk. The redness appears to belong *on* the mug, and when I move the mug the redness follows it. And yet we know that the neuroanatomical structures involved in generating these obviously connected perceptions are located in different places in the brain. The brain regions that recognized the mug are distinct from those that located it as being on my desk. Likewise, the brain regions that perceived its redness are distinct from the ones that saw it move. And so on. The problem of how all this information *comes together* to form an ordinary unified experience of consciousness is called the binding problem.

One approach to solving this problem has been to attempt to identify anatomical structures that receive input from *all* the different perceptual processing modules. Stein and Meredith's book, *The Merging of the Senses* (1993)—although a little technical—provides a comprehensive survey of these structures. We pointed out in chapter 1 that the association cortex, at the point where the occipital, temporal, and parietal lobes *converge* to form transmodal "directories," is one such structure. We also mentioned the superordinate connections of the *prefrontal lobes* in this context.

Wolf Singer's group (Engel et al., 1991; Gray et al., 1989; Gray & Singer, 1989) championed an entirely different approach to the binding problem, usually referred to as the *"forty-hertz" hypothesis*. This group demonstrated that posterior cortical cells seem to fire in *synchrony* during a conscious visual experience, roughly at a 40-hertz rhythm of oscillation. They argued that each brief moment in which these neurons fire simultaneously represents a unified *unit of consciousness*. The unity of consciousness is thereby forged by linking neuronal activity together in time rather than place. (Every second of consciousness is made up of 40 "micromoments" that occur so quickly that consciousness is experienced as continuous.)

Both of these approaches have merit and go some way toward solving the binding problem. However, neither dispenses entirely with the homunculus problem. Binding bits of consciousness in place merely seems to *localize* the homunculus; binding them in time still begs the original question: *What perceives* these synchronized neuronal oscillations?

There is a third approach to the problem, which we favor. This approach combines aspects of both of the approaches just described. We outline this approach in detail in chapter 3, but, briefly stated, it suggests that what binds our external perceptions together is the fact that they are grounded in our *internal* perceptions—which are, in turn, perceptions of our *bodily* selves. It is the fact that each of us exists in a singular body, therefore, that ultimately binds our consciousness together. But before we can explain why the visceral body is the bedrock of consciousness, we need briefly to introduce another basic question.

WHAT USE IS CONSCIOUSNESS?

Although neuroscientists have begun to address the question of *how* and *where* the brain generates consciousness, the question remains as to *why* consciousness exists in the first place. Another way of putting this question is to ask: What *evolutionary advantage* does consciousness bestow?

We have already seen that *intelligent behavior* does not require consciousness (assuming that computers are unconscious). Investigations across a broad range of human neurological patients have shown the same thing. Cognitive abilities are retained to a remarkable extent in the absence of conscious experience. In chapter 3 we demonstrate this in detail when we discuss the phenomena of "*blindsight*" and *unconscious memory*. Philosophers call people who perform intellectual feats in the absence of consciousness "zombies." This is an unfortunate choice of phrase in relation to clinical patients, but such people really do exist and, in fact, in quite large numbers. This presents an intriguing problem. If it is possible for people and machines to behave intelligently (to solve problems, adapt to prevailing conditions, learn from experience, etc.) then what is consciousness *for?* If we can behave intelligently without it, then why do we need it? In chapter 3, we argue that the core function of consciousness, too, is linked to the fact that each of us exists in a *body*.

We must now enter into a detailed consideration of these deferred scientific issues. Before doing so, we conclude this chapter with some final conceptual points.

SO, WHAT *IS* "THE MIND"?

We have arrived at the following provisional conclusions. The mind itself is unconscious, but we perceive it consciously by looking inwards. It is this capacity for "looking inwards" (for introspection or self-awareness) that is the most essential property of a mind. The "me" that we perceive through introspection can *also* be perceived through our external senses as a physical body (made of organs, or physiological processes). The body is not the mind. Bodily processes are not intrinsically mental; they can even be performed by machines. That is why we say that the mind *itself* is unconscious. It is only our *perceptions* of these underlying processes that are conscious—and, what is more, these perceptions come in two forms, only one of which is described as "mental." The mind, therefore, is intimately bound up with the first-person observational perspective. This is the only perspective from which everything we observe can be grounded in a background sense of self, which is ultimately generated by our inner awareness of living in a physical body. A computer can only be rendered conscious if it is imbued with this capacity for self-awareness, grounded in a visceral body.

SOME METHODOLOGICAL ISSUES, REVISITED

We argued at the end of chapter 1 that it is advantageous for scientists to study *matter*, the aspect of the world that is experienced by our external senses, because there are things one can do experimentally with matter (brains, for instance) that one cannot do with subjective experiences. It should now be a little clearer what we meant by that. Only matter—or only external perception—permits the reliability of *multiple observers*. This reduces the potential for *observer bias* that occurs with lone observers (like psychotherapists). Minds, by definition, cannot be observed by multiple observers. Similarly, from the external point of view we can study the mental instruments of *nonhuman animals*. Nonhuman animals cannot provide verbal reports of their experiences. However, as we shall see in the ensuing chapters, neuroscience has learned a great deal about the functions of the human brain (and therefore about the functional organization of the mind) by generalizing what has been learned from other animals, particularly other mammals. The relevant experiments could never have been performed on humans.[11]

These two examples (which we could easily multiply) demonstrate the final point we want to make in this chapter, which is simply this: If we accept the argument presented in this chapter—namely, that the mental apparatus is observable from two different perspectives simultaneously, firstly as a material object and secondly as subjective awareness—then it is obvious that we have everything to gain, and nothing to lose, by integrating our two modes of inquiry. As with the allegory of the blind men and the elephant, the true nature of the mental apparatus will only become apparent when our dual perspectives on it converge on a single set of conclusions. Conclusions derived from subjective data (from the psychoanalytic method, for example) provide only one kind of evidence. And this evidence has definite limitations from the scientific point of view. On the other hand, subjective data should not be despised; *it provides evidence that cannot be gained from any other perspective.* It should therefore be clear why we believe that such an integration does not *reduce* the one perspective to the other, and also why we said at the end of chapter 1 that linking the invisible world of subjectivity with the visible tissues of the brain deepens immeasurably what we can discern with our "objective" scientific eyes.

[11]The ethical question as to whether or not these experiments *should* ever have been performed on nonhuman animals is another matter, which we address briefly in chapter 4.

CHAPTER 3

CONSCIOUSNESS AND THE UNCONSCIOUS

The next few chapters each address different aspects of mental life that have received substantial neuroscientific attention in recent years. We begin with the most general of them—consciousness—and thereby continue where we left off at the end of the previous chapter.

A QUIET REVOLUTION

Freud was one of the first to claim (over a hundred years ago) that most of our mental life operates *unconsciously* and that consciousness is merely a *property* of *one part* of the mind. To hold this opinion in medical science at that time was highly controversial. Much else that Freud proposed all those years ago is *still* hotly contested. However, the notion that most mental functioning operates unconsciously is very widely accepted in cognitive neuroscience today. One of Freud's most fundamental innovations has thus entered the mainstream of contemporary science. This does not mean that modern neuroscientists accept everything that Freud said about the unconscious in the *psychoanalytic* sense. But that is another matter, which we address later. To begin with, we shall limit our consideration of the brain mechanisms of consciousness and unconscious mental activity to the purely descriptive meanings of those terms.

Many cognitive scientists now argue that consciousness is of very *little* importance in mental life and that the vast bulk of our mental operations are performed without consciousness (for a review see Bargh & Chartrand, 1999). Given the fact that mainstream scientists have frequently been hostile to psychoanalysis, this reversal of opinion is a remarkable development. There are many reasons for this change. Various lines of evidence have convinced neuroscientists that Freud must have been right on this point. The most dramatic evidence derives from the clinical observation of brain-damaged patients.

BLINDSIGHT

The term **blindsight** (Weiskrantz, 1986) is applied to patients with damage to the visual cortex of the occipital lobes—the primary visual cortex—which is where most of the nerve fibers from the retinae terminate (see chapter 1). Such patients suffer from "cortical blindness"; they are blind because the part of the cortex that generates visual consciousness is no longer working.[1] Blindness, then, means a lack of visual *awareness*. Thus, if you were to hold an object before these patients' eyes and ask them what they see, they would respond with the obvious: "I don't see anything; I'm blind." But when they respond in this way, they are actually mistaken. They are erroneously equating "*seeing*" with "seeing *consciously*." The distinction between vision and

[1] Throughout this book, for convenience's sake, we will use loose phrases like "the part of the cortex that generates visual consciousness." What we mean by such statements is that activation of the part of the brain in question generates neural activity that is the physiological correlate of the type of consciousness in question. (See chapter 2.)

conscious vision is demonstrated when you ask these same patients to make a "forced choice" between various options (in other words, you encourage them to *guess*). The results of such experiments reveal that they guess correctly at a level well above chance, which demonstrates that these patients are seeing—are processing visual information—without *realizing* it (see Weiskrantz, 1986). They are seeing *unconsciously*. This occurs because some visual information is projected from the retina onto other parts of the cortex (intact in these patients) that do not generate visual *consciousness* but are, nevertheless, equipped to process the visual *information* they receive. In other words, these patients—as far as visual information is concerned—act like the "zombies" mentioned in chapter 2. Their brains *compute* visually, but they do not possess visual consciousness.

IMPLICIT MEMORY

The same thing occurs with respect to other cognitive faculties. It is not all that rare for neurological patients to lose the ability to lay down new memories. This condition is called *amnesia*. These patients remember (recall consciously) nothing that happens to them after the onset of their brain disease or injury (see chapter 5). If you were to read a list of words to such patients, after a few minutes they would not only forget the words, they would even forget the fact that you read them the list. However, as with cases of cortical blindness, such patients can be encouraged to "guess," using the forced-choice paradigm. When they do so, they "randomly" select or generate words that were on the original list, at a very much higher rate than chance. So, just as we can *see* unconsciously, we can also *remember* unconsciously. The technical term for this unconscious type of remembering is **implicit** memory (conscious remembering is **explicit** memory).

SPLIT-BRAIN STUDIES

In so-called **split-brain** patients, to treat otherwise intractable epilepsy the *corpus callosum* (see chapter 1) has been severed, thus separating the left (language-dominant) hemisphere of the brain from the right (see chapter 8).

By briefly flashing an image on a screen to such patients, it is possible to provide the right hemisphere with information that the left hemisphere cannot access. On this basis, it is possible to influence the patient's behavior without him or her being consciously aware of it. In one of Nobel Prize-winning neuroscientist Roger Sperry's famous cases, pornographic pictures were projected to the isolated right hemisphere of a patient. The patient blushed and giggled. When Sperry asked her why she was embarrassed, she was unable to account for it. This case (described in Galin, 1974, p. 573) demonstrates that an entire cerebral hemisphere can process information "unconsciously."

The case also reveals something else that is of crucial importance for understanding consciousness. The visual cortex was completely intact in Sperry's patient. This means that the pornographic pictures were perceived by the visual-consciousness-generating part of her right hemisphere. Why, then, did she appear not to know what she saw? The answer to this question provides a good illustration of the "functional systems" concept we discussed in chapter 2. Although it is true that the primary visual cortex (in either hemisphere) is capable of generating simple visual consciousness, it does not do so *in isolation*. For someone to *reflect* consciously on visual experiences, he or she has to recode the visual experiences into *words*. This capacity is lost when the left (verbal) hemisphere is disconnected from the original visual experience. This shows that a distinction needs to be drawn between two levels or types of consciousness: **simple awareness** and **reflexive awareness**. It also shows that the function of reflexive awareness is intimately connected with the left cerebral hemisphere and therefore with *words* (or, rather, "inner speech"). We discuss these complex issues again, in a little more detail, later in this chapter (and also, in greater detail, in chapter 8).

The fact that an entire hemisphere (about half of the forebrain) can, in a sense, function unconsciously raises an intriguing question.

HOW MUCH OF MENTAL LIFE IS CONSCIOUS?

There are various ways of addressing the question of how much of mental life is conscious, each of which leads to slightly different answers. What they all reveal, however, is that *consciousness is a very limited part of the mind.* For example, if the extent of consciousness is equated with the amount of information that we can "hold in mind" at any one point in time, then readers might be surprised to learn that consciousness is restricted to only *seven units of information.* It is no accident that most telephone numbers are roughly seven digits long! Digit span (the capacity to repeat a string of random digits) is a standard clinical test of an aspect of **working-memory** capacity. ("Working memory" is synonymous with the ability to consciously "hold things in mind"; see chapter 5). If a patient cannot retain roughly seven digits, his or her audioverbal working memory (audioverbal consciousness) is considered abnormal. We appear to hold visuospatial information (or "location" information) in mind in a similar way, but this aspect of consciousness is even more restricted: most people can hold only some *four* units of visuospatial information in mind at a time. (This capacity is usually tested by tapping a series of blocks scattered before the patient and asking him or her to hold the sequence of taps in mind.) Considering how many thousands of pieces of information we are processing all the time, this way of measuring the capacity of consciousness reveals that it is very limited indeed. The vast bulk of the information we constantly need to process must be processed in the unconscious part of the mind.

Another way of estimating the "size" of consciousness is to measure the extent of its *influence on our behavior*. What proportion of our actions is consciously determined? In a review of the scientific evidence pertaining to this question (and related matters), Bargh and Chartrand (1999) concluded that 95% of our actions are unconsciously determined. This way of measuring consciousness therefore suggests that it accounts for only 5% of our behavior.

So, regardless of how they measure it, mainstream cognitive scientists today agree with Freud on this point: consciousness is attached to only a very small part of our mental life. *Where*, then, is this consciousness generated in the brain? And *how* does it become attached to mental processes? And *why*?

THE CORTEX: CONTENTS (OR CHANNELS) OF CONSCIOUSNESS

It was once undisputed that the *cortex* is the seat of consciousness. This is because damage to different parts of the cortex so obviously deprives patients of the different perceptual modalities that channel through them. Thus, visual consciousness is dependent on visual cortex (in the occipital lobe), auditory consciousness on auditory cortex (in the temporal lobe), and so forth (see chapter 1). Consciousness was therefore traditionally attributed to certain core regions of the cortical unit for receiving, analyzing, and storing information, as discussed in chapter 1. Consciousness is not attributed to the peripheral sensory organs themselves, for various reasons. First, these organs are intact in cases of cortical blindness, cortical deafness, and the like.

Second, and more importantly, patients who *acquire* a peripheral sensory loss (i.e., those who are not *born* blind or deaf, etc.) retain conscious *mental imagery* in the affected modality. Thus, for example, peripherally blind people are still capable of experiencing visual dreams. Accordingly, direct stimulation of modality-specific cortex generates conscious sensations in the relevant modality, even if the peripheral sensory organ that projects to it is destroyed completely.

We have said already that visual (and auditory, etc.) *consciousness* is not synonymous with visual *processing.* Patients with blindsight can "see" unconsciously precisely because not all cortical visual processing is conscious. On this basis, neuroscientists have identified even more precisely the regions of cortex that generate the various modalities of conscious awareness. We have also learned (see above) that for patients to *reflect* upon the simple experiences generated by these unimodal cortical areas (i.e., to turn experience into awareness *of* experience), other, mainly language-based mechanisms are required. These other cortical mechanisms are responsible for generating reflexive (as opposed to simple) consciousness.[2] Consciousness therefore also involves an important contribution by the functional unit for programming, regulating, and verifying activity, as discussed in chapter 1.

The problem of *where* consciousness is generated in the brain therefore appears to be quite easily solved. But even the "easy problem" is not *that* simple. The conclusions we have summarized so far all derive from a research tradition that reduces the contents of consciousness to **qualia** derived from

[2] In chapter 2, we alluded to the fact that such mechanisms might also play a part in "binding" the different modalities of simple awareness together to form more complex (multimodal) experiences. But this solution to the binding problem still left us with the homuncular problem.

external perception—colors, sounds, and so on—or combinations of such qualia, and, perhaps,[3] abstractions derived from them. However, a second approach to the neurology of consciousness also exists. The two traditions were, surprisingly, integrated only very recently. Whereas the cortical tradition focuses on the *contents* (or perceptual qualities) of consciousness, the second tradition focuses its research efforts on the *level* (or *state*) of consciousness (these terms were introduced in chapter 1).

THE BRAINSTEM: LEVELS (OR STATES) OF CONSCIOUSNESS

This state aspect of consciousness is the one with which anesthesiologists are most concerned. The same applies to the families of, say, road-traffic-accident victims, whose loved ones are *unconscious* in the sense of having slipped into states of coma. "Consciousness" in this context refers to the *global state of being awake, aware, and alert*. The state of consciousness is a background level of awareness—a "global workspace" (Newman & Baars, 1993) within which its more specific *contents* take place. It is like a page upon which the contents of consciousness can be written. This aspect of consciousness is normally described in *quantitative* rather than qualitative terms. In clinical situations, the level of consciousness is graded on a 15-point scale (the Glasgow Coma Scale). Loss of this aspect of consciousness following road-traffic accidents (and the like) is not caused by widespread brain damage that diffusely affects all the cortical

[3] We say "perhaps" because it is unclear whether or not our abstract thinking occurs in a concrete audioverbal medium (cf. "inner speech"). See Baars and McGovern (1999) for a discussion of this issue.

structures discussed in the previous section. On the contrary, only a very specific, and very small, region of the brain is involved. And this region is not in the cortex at all.

There is extensive evidence that certain structures in the *brainstem* are critical for generating the *global state* of consciousness. A group of structures running up the core of the brain, above the medulla oblongata through the pons, extending upward through the midbrain into part of the thalamus, is particularly implicated here (see Figure 3.1). These tightly interconnected nuclei were initially not recognized as being separate, and for that reason they were originally called the **reticular activating system** (a "reticulum" is a continuous network). We now distinguish a number of nuclei within this system, which still includes the classical reticular formation. The "reticular" activating system was discovered in the 1950s by Giuseppe Moruzzi and Horace Magoun. Today, we call this system the **ascending activating system** or the **extended reticular and thalamic activating system** (ERTAS).

The brainstem is roughly the size of the human thumb, and the nuclei in question are roughly the size of match heads. It is a remarkable fact that damage to such a tiny region of the brain leads to an absolute obliteration of consciousness, producing a

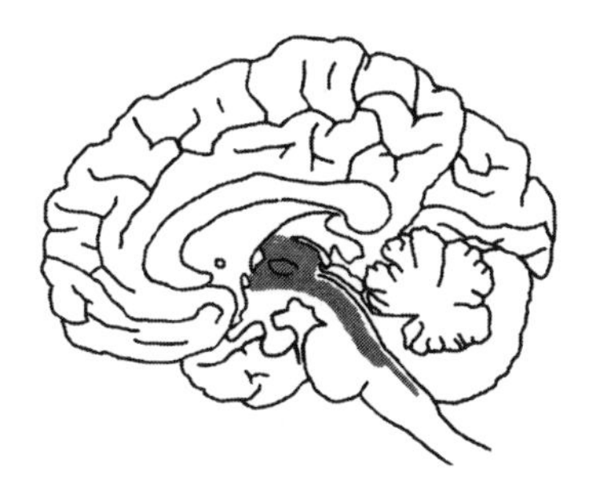

FIGURE 3.1
The ERTAS

deep coma. General anesthesia rests on little more than various ways of modifying the output of this tiny system. We might well say, then, that *these* tiny nuclei are the seat of consciousness. On this view, consciousness is generated not by specific cortical zones but, rather, by the *activation* of specific cortical zones by these deep structures. Many of the complexities of conscious versus unconscious information processing can then be explained in terms of the selective directing of attention (i.e., core brain activation) to and from the cortical zones in question.

WHAT IS CONSCIOUS "STATE" A PERCEPTION OF?

We owe the integration of these two lines of research on the neural basis of consciousness to the pioneering work of a neurologist named Antonio Damasio. In a recent book, *The Feeling of What Happens* (1999b), Damasio asked this question: If the qualia of consciousness are derived from external perceptual mechanisms, what is the equivalent derivation of the quantitative variety of consciousness? We know that the content of consciousness represents the pattern of activation of cortical tissues responsive to changes in the outside world; does the level or state of consciousness generated by the deep brainstem *represent* anything, or *mean* anything?

The ERTAS includes certain thalamic nuclei, parts of the hypothalamus, the ventral tegmental area, the parabrachial nuclei, the periaqueductal gray, the nucleus locus coeruleus, the raphe nuclei, and the reticular formation proper (Figure 3.2). Most of these structures were mentioned in chapter 1, where they were described as core components of the functional unit for modulating cortical tone and arousal.

We also mentioned in chapter 1 that these nuclei contain the source cells for neurotransmitter systems that project widely

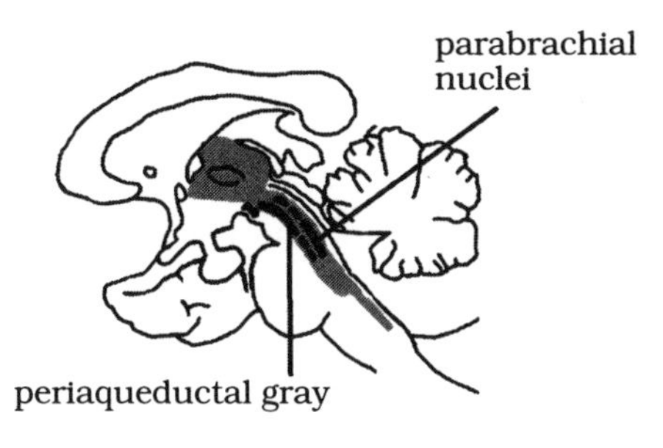

FIGURE 3.2
Some nuclei comprising the ERTAS

in the forebrain, and they are of great interest not only to anesthesiologists but also to psychiatrists. The reason why psychiatrists are so interested in these nuclei (each of which has a different set of functions) is made clear in the next few pages and is further elaborated on in chapter 4, where we discuss emotion and motivation. Some of the main neurotransmitters excreted by these cells are dopamine, serotonin, norepinephrine, histamine, and acetylcholine (see chapter 1).

TWO SOURCES OF INFORMATION, REVISITED

Damasio's search for the *source input* of these "state-dependent" cells of the brainstem led him into a vast field of knowledge that had previously been regarded as holding little interest for mental science. All the core brain nuclei listed above are centrally involved in the *modulation and regulation of our visceral states*: temperature regulation, glucose (blood sugar) metabolism, and the like. Their source input is information about the state of the internal milieu, delivered not only via classical neurotransmitter systems, but also directly via hormones that course through the bloodstream and cerebrospinal fluid circulation (see chapter 1). On this basis, Damasio (1999b) reached the simple conclusion that whereas the "content" of consciousness is attached to the posterior cortical channels that monitor the outside world, the "state" of consciousness is a product of the ascending activating system of the brainstem, which monitors the internal milieu of the body. Thus, whereas the contents of consciousness represent changes in cortical zones derived from one's external perceptual modalities, the *state* of consciousness represents changes in the internal situation of one's body.

Moreover, just as the association zones of the posterior cortex not only receive and analyze external perceptual information but also *store* it, so too these deeper, inwardly directed networks contain representational "maps" of our visceral functions. For this reason, just as the contents of consciousness reflect not only *concrete* changes in the external world, but also *thought* activity (internally generated images), so too fluctuations in the state of consciousness are responsive not only to *actual* visceral events (e.g., falling core body temperature or rising blood sugar) but also to changes in the networks that *represent* these functions, whatever the source of those changes might be. Conscious state is generated by a *virtual* body. Furthermore, as we pointed out in chapter 1, it is important to remember that these structures do not only *perceive* information derived from the external and internal worlds, respectively, they also *act upon* that information and thereby modify its sources.

So the background "state" of consciousness *does* mean or represent something. It represents "you"—the most basic *embodiment* of your "self." More than that, it represents the current *state* of your self: "This is me, I am this body, and, right now, *I feel like this*." Far from being without quality, the background state of consciousness is therefore *replete* with meaning and feeling—indeed, it is the very bedrock of personal meaning and feeling. This aspect of consciousness therefore not only "represents" your self, it also tells you how you are doing.

THE FUNCTION OF CONSCIOUSNESS: INTEGRATING THE TWO WORLDS

All of a sudden, the function of consciousness—which seemed so elusive a few pages ago—falls into place. How, without consciousness, would you know how you feel? *That* is the function of consciousness. It is not only intrinsically introspective (as we said in chapter 2), it is also intrinsically *evaluative*. It imparts *value*. It tells us whether something is "good" or "bad"; and it does that by making things *feel* good or bad (or somewhere in between). That is what consciousness, feeling, is *for*. (And that is why psychiatrists are interested in modifying the chemical outputs of these core brainstem nuclei.)

The evaluative function of our conscious "state" has its roots in the visceral monitoring structures of the core brain. This function of consciousness is therefore intrinsically biological. Its evolutionary survival value is obvious: How long would we survive if we did not have a way of monitoring the delicate economy of the internal milieu of our bodies? As we mentioned in chapter 1, the organ systems of our bodies can only function effectively within a very narrow range of set-points—with respect to temperature, blood-sugar level, and so forth. The most basic function of consciousness, then, is to monitor the state of these homeostatic systems and to report whether they (i.e., you) are "contented" or not.

But bodily self-monitoring is only the *most basic* function of consciousness. All our vital inner needs can only be met in the external world. The inner state of consciousness (which tells us, above all, what our current needs are) therefore has to be brought into connection with the current state of the world around us. Although, as we have seen, it is not *necessary* to be conscious of the external environment in order to perceive it, it is nevertheless *useful*. It is useful to be able to say things like, "I feel like *this* (hungry), so I want to eat *that thing* over there," or, "I feel like *this* (upset), because *that thing* over there bit me." In this way, consciousness—that is, *value*—is imparted to *objects*, and objects come to be known as "good" or "bad." Consciousness is not only what you feel, it is what you feel *about* something.

Thus, even if the evolutionary "dawn of consciousness" was purely introspective, in a rudimentary biological sense, it probably quickly generalized, and our external perceptual modalities, too, became imbued with feeling (with consciousness). In this way, our external perception was transformed from being a set of (unconscious) information-processing channels into being the generator of the rich texture of perceptual *qualities* (conscious sights, sounds, smells, etc.) that we are now able to experience. This is consistent with the anatomical fact that the output of the core brain nuclei in question is broadcast very widely to the forebrain, and with the physiological fact that such "bottom-up" activation is necessary before higher cortical processes can become conscious.

Damasio (1999b) therefore concluded that consciousness consists of more than mere awareness of our inner states; rather, it consists of fluctuating *couplings* of the current state of the self with the current state of the object world. Each unit of consciousness forges a link between the self and objects. These momentary "units" of conscious time are probably generated by the rhythmical oscillations mentioned in chapter 2 (e.g., the 40-hertz oscillations that characterize visual awareness). These oscillations are generated by pulses of activation of cortex, emanating from deep "reticulate" thalamic nuclei, thereby coupling the two varieties (or sources) of consciousness with one another many times per second. This is how we generate "the feeling of what happens" that provided the title of Damasio's book. Consciousness thus consists of feelings (evaluations) projected onto what is happening around us. Or, to put it the other way round, consciousness consists of awareness of what is happening around us, grounded in a background medium of self-awareness. Note especially that this explanation of consciousness solves both the binding problem and the homuncular problem (discussed in chapter 2). The various "channels" of consciousness are bound together by the grounding "state" of consciousness, which is itself the homunculus; the little person in your head is literally a projection of your bodily self.

Damasio calls this coupling mechanism "**core consciousness**." Some further complications of consciousness are to be discussed in a moment. First we want to briefly mention two interesting asides.

DAMASIO AND FREUD

The previous chapter offered a short précis of Freud's version of the two perceptual surfaces of consciousness (Freud, 1940a [1938]). The similarity between Freud's model and Damasio's, just described, is very striking. When one of us [MS] pointed out this deep compatibility to Damasio, in a piece in a published commentary he stated: "I believe we can say that Freud's insights on the nature of consciousness are consonant with the most advanced contemporary neuroscience views" (Damasio, 1999a, p. 38; see also Crick & Koch, 2000). There are also many points of contact between Damasio's neuroscientific theory and those of other psychoanalytic theorists. Here, then, are some rich veins for future collaborative research between the two disciplines (see chapter 10).

CONSCIOUSNESS IN MACHINES AND IN NONHUMAN MAMMALS

When one starts thinking about the problem of consciousness in the way that Damasio suggests, the question of whether or not a machine can be conscious begins to appear rather ridiculous. Some day this question might only be asked by people who are unfamiliar with the essential neuroscientific facts about consciousness. Consciousness has everything to do with being *embodied*, with awareness of one's bodily state in relation to what is going on around one. Moreover, this mechanism seems to have evolved only because bodies have *needs*. Consciousness is therefore deeply rooted in a set of ancient biological *values*. These values are what feelings *are*, and consciousness *is* feeling. It is therefore very difficult to imagine how, why, and where a disembodied machine would generate consciousness. This does not rule out the possibility of an artificial system with self-monitoring properties. But the *self* that it monitors would have to be *a body* (and preferably one with a long evolutionary history) if it is really going to generate *feelings*.

This argument has interesting implications, too, for the question of consciousness in other animals. It suggests that any animal with a brainstem designed roughly like our own—that is, a brainstem that modulates visceral processes and relays its output to cortex—is likely to experience consciousness. As it happens, *all mammals* have brainstems with nuclei that are structured and connected in roughly the same way as are those of humans—their brainstem nuclei even excrete the same chemicals (and deliver them to roughly the same places) as their human counterparts. There is therefore very good reason to believe that dogs, cats, dolphins, whales—even laboratory rats and mice—possess "core consciousness." This implies that all mammals share our most basic (biologically rooted) values. The same elementary things are likely to make a mouse and a human being feel "good" and "bad." In chapter 4 we shall learn that mice, no less than men, most probably feel pleasurable excitement when anticipating the consummation of a need, fear in the presence of an enemy, anger when prevented from having what they want, distress on being separated from loved ones, and so on. Recognizing these facts has profound ethical implications for humanity.[4]

There are, however, "higher" levels of consciousness which involve neural structures that we do *not* share with other mammals. As a result, the nature of the *cognitive aspects* of their consciousness is likely to differ quite dramatically from our experience.

EXTENDED CONSCIOUSNESS

Damasio subsumes these higher cognitive aspects of consciousness under the heading of "**extended consciousness**." Just about every consciousness theorist divides consciousness in some way along these lines. What Damasio calls core consciousness is roughly equivalent to what other theorists call **simple** or **primary consciousness**, and what Damasio calls extended consciousness is similar to what is often called **reflexive** or **secondary consciousness**. These latter terms all refer to "consciousness of consciousness." That is, they denote not only awareness of what you feel now, but awareness *that* you feel that now.

This aspect of consciousness is not limited to simple *perception*; it also involves *thinking* about perception (or thinking *with* perceptual images). This is also not limited to *present* perceptions; it is possible to think about (or with) the residues of *past* perceptions. Although these functions of extended conscious-

[4] It even holds out the possibility that answers to the philosophical problem of how to live a worthwhile and fulfilling life might someday be grounded in objective, biological facts.

ness are probably not unique to humans, they are certainly far more highly developed in humans than in other mammals, even in comparison to our nearest primate relatives (see chapter 9).

Extended consciousness depends heavily on *cortex*, and on *association* cortex in particular. It depends primarily on the functional contribution of the *language zones of the left cerebral hemisphere* and—above all—on the superstructure of the *prefrontal lobes*. This brain region is vastly more developed in humans than in other mammals. It forms the unit for the programming, regulation, and verification of action, as described in chapter 1, and it therefore has the capacity to *re-represent* the units of core consciousness originally represented (perceived and stored) in the posterior cortical and paralimbic zones. This allows us to reflect on, think about, and remember our conscious experiences, as opposed to simply living them moment-by-moment.

We have said that the ability of humans to be conscious of their consciousness, and especially to transform concrete perceptions into abstract concepts, is heavily dependent on our capacity for *language*. Language enables us to activate the perceptual trace not only of one *particular* object (e.g., the visual image of one's father) but also of a whole *class* of objects (audio-verbal traces of *words* like "fathers" or "women"). Moreover, it permits us to think consciously about the *relations between* concrete things, using function words (like "my father *loves* me") and abstractions (like "he is bigger, older and wiser *than* me").

EXTENDED CONSCIOUSNESS AND MEMORY

Extended consciousness also spreads consciousness over time. The "feeling of what happens" is always influenced by the feeling of what happened *previously*. For example, when core consciousness generates the momentary state of awareness, "I am reading this book," it carries with it the memory of what you read a moment ago (at the beginning of this sentence). This capacity, which enables you to make sense of what you are reading while you are reading it, depends on an aspect of extended consciousness known as "**working memory**" (which we mentioned earlier in this chapter). The experience, "I am reading this book," also carries with it a host of implicit knowledge, derived from a lifetime of being you and reading books in general. These are "**procedural**" and "**semantic**" aspects of memory, which are not normally conscious. But it is possible to bring particular instances of such previous experiences to explicit awareness (e.g., the memory of another book on a similar topic that you read a few years ago). Retrieval of such memories depends on an aspect of extended consciousness that neuropsychologists call "**episodic memory**" (memories of previous instances of the self in relation to objects). This type of memory is lost with lesions to the hippocampus, which causes the type of amnesia we discussed earlier in this chapter. (These topics are discussed in more detail in chapter 5.)

Access to this rich array of memories allows for the development of what Damasio (1999b) calls an "**autobiographical self**." This aspect of the "sense of self" is built upon, but greatly extends, the fleeting awareness of "self" that constitutes the grounding medium of core consciousness. In psychoanalytic terms, the core "self" might be described as a perception of the current state of the "id," whereas the extended, autobiographical "self" is synonymous with the "ego." The "autobiographical self" is dependent on past experience, but this manifestation of extended consciousness also makes it possible to imagine (and plan for) the *future*. This (feedforward) aspect of extended consciousness, too, is intimately bound up with the functions of the prefrontal lobes.

A final point on the matter of extended consciousness relates to *hierarchy*. If core consciousness is disrupted, extended consciousness will necessarily be lost. We understand this from studies of coma, anesthesia, and certain types of epilepsy. Although core consciousness is therefore a prerequisite of extended consciousness, the reverse is not true. It is possible to destroy aspects of extended consciousness without disrupting core consciousness. Under such circumstances, extended consciousness will be "distorted" in some way, as the remaining systems try to cope with the absence of an important psychological process. There are a number of ways this can happen, because extended consciousness employs a wide range of higher cognitive processes, distributed across a large number of brain regions. The fact that *core* consciousness remains intact with focal damage to these higher regions reinforces its status as the *fundamental* basis of consciousness.

THE UNCONSCIOUS

If we were to ablate *all* the structures that are the neural correlates of consciousness, what would we be left with? In a purely *descriptive* sense, what we would be left with is "the unconscious." But this "unconscious" would not behave anything like *the* Unconscious (with a capital U) of Freudian psychology. We would not be left with a seething cauldron of instinctual impulses. In *real* cases where consciousness is obliterated completely, what we are left with is a person in a state of coma, with no evidence of mental life of the conscious *or* unconscious variety. This is partly due to the fact that in such cases the loss of consciousness is actually attributable to the destruction of core brainstem nuclei that perform functions that Freud would have attributed to the "id." Their lack of mental activity is therefore attributable to a lack of "drive" (see chapter 1) and explicable according to the hierarchical arrangement just outlined.

But is there a part of the brain that embodies the Freudian system "Conscious" which, when damaged, releases the functions that he attributed to the system "Unconscious"? There certainly is, but before we can describe it, we need to remind ourselves of some of the basics of Freudian theory.

A HISTORICAL ASIDE ON THE DYNAMIC UNCONSCIOUS

It is important to remember that Freud *abandoned* his original idea that the functions of the mind should be divided between the systems Conscious (or Conscious and Preconscious: *Cs–Pcs*)[5] and Unconscious (*Ucs*). In 1923, he recognized that the rational, reality-constrained, executive part of the mind is not necessarily conscious, and not even necessarily *capable* of becoming conscious (Freud, 1923b). Consciousness, for Freud, was therefore not a fundamental organizing principle of the functional architecture of the mind. Accordingly, from 1923 onward, Freud redrew his map of the mind (see Figure 3.3) and attributed the functional properties previously attributed to the "system *Cs-Pcs*" to the "ego"—where only a small portion of the ego's activities were conscious (or capable of consciousness). The ego was mainly unconscious. Its core functional property was the capacity, not for consciousness, but, rather, for *inhibition*. Freud considered this capacity (the capacity to inhibit drive energies) to be the basis of all the ego's rational, reality-constrained, and executive functions. This inhibitory capacity was the basis of what Freud called "secondary-process" thinking, which he contrasted

[5]Preconscious, in Freud's terminology, means capable of becoming conscious. (In Figure 3.3, "Pcpt." stands for "perception.")

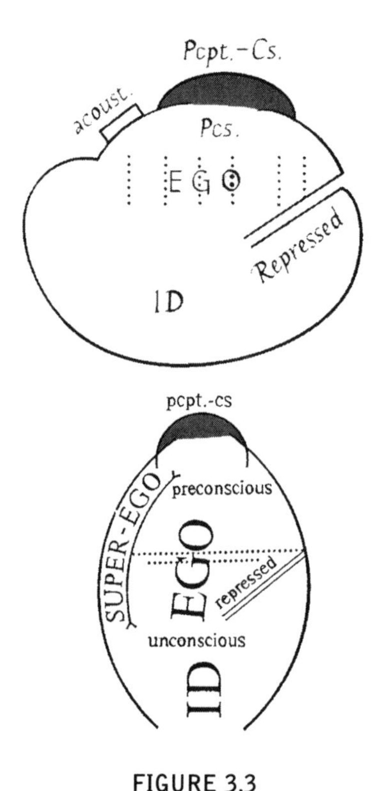

FIGURE 3.3
Freud's structural model of the mind
(*top*, from Freud, 1923b, p. 24; *bottom*, from Freud, 1933a, p. 78)

with the unconstrained mental activity that characterized the "primary process." It was this property (rather than consciousness) that gave Freud's ego—the "autobiographical self" of Damasio—executive control over the otherwise automatic, biologically determined functions of the mind.

So, when we look for a part of the brain that, when damaged, releases the functions that Freud attributed to the "system Unconscious," what we are actually looking for is (1) a brain region that is not necessarily implicated (or not centrally implicated) in the consciousness-generating functions of the mind, but (2) *is* centrally implicated in the *inhibitory* functions of the mind.

PHINEAS GAGE, REVISITED

The *ventromesial quadrant of the frontal lobes,* more than any other brain region, meets these criteria (Figure 3.4). Bilateral damage to this part of the brain does indeed produce a state of mind that shows several properties reminiscent of what Freud (1915e) described as "the special characteristics of the system *Ucs.*" These functional characteristics were listed as follows: "exemption from mutual contradiction, primary process (mobility of cathexis), timelessness, and replacement of external by psychical reality" (p. 187). Phineas Gage (see chapter 1) sustained *unilateral* damage in the ventromesial quadrant of his left frontal lobe, and therefore he displayed some of these characteristics. A number of more severe cases, with *bilateral* damage in this area, were described recently by Kaplan-Solms and Solms (2000).

Exemption from mutual contradiction One of their cases involved an English gentleman in a neurological rehabilitation unit, who had lived abroad for some years. A close friend of his had died some 20 or 30 years previously, while they were both living in Kenya. One day the patient informed the staff excitedly that he

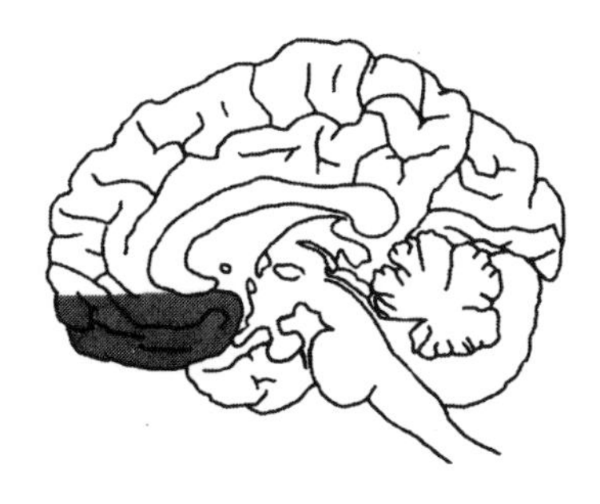

FIGURE 3.4
The ventromesial quadrant of the frontal lobes

had met a friend of his in the hospital. "Can you believe it," he said, "Phil Adams[6] is here in the same hospital as me. You know, the chap I told you about who died in Kenya 20 years ago; it is wonderful to see him again." When questioned as to how Phil Adams could be in the hospital if he had died in Africa 20 years before, the patient stopped for a moment and said: "Yes, that must cause interesting legal problems—being dead in one country and alive in another." This man was clearly quite capable of accepting two mutually exclusive facts as being simultaneously true.

Timelessness A second patient—a woman who suffered from damage in the same brain region—had experienced several instances of medical difficulties prior to the stroke for which she had been admitted on this occasion. One was a deep-vein thrombosis (in her leg), another a hysterectomy. To this woman, her current hospitalization was one and the same with the others. She would speak as if she were in the neurological ward for the purposes of a hysterectomy, but in virtually the next sentence she would suggest that her admission was due to a deep vein thrombosis, and then again, also, for a stroke. Indeed, she even seemed to think that she was hospitalized at all the locations of the previous admissions simultaneously—so that she was in King's College Hospital, the Royal Free Hospital, and the Royal London Hospital, all at the same time. A series of separate temporal events had thus become merged into a single experience.

Timelessness of a different kind was displayed by the gentleman with the dead friend, described above. His wife always came to see him at 5 p.m., which was visiting time. For this reason, the patient was constantly of the opinion that it was 5 p.m.—

[6] A pseudonym.

even straight after breakfast, or before lunch. Following one breakfast, when his error was being corrected by a staff-member for the umpteenth time, he noticed a "NO SMOKING" sign on the wall, which took the form of a red circle with a diagonal line through it. Mistaking this sign for a clock, he retorted: "Look . . . it *is* 5 o'clock!"

Replacement of external reality by psychic reality In cases such as those just presented, the demands of the internal world of the drives take precedence over the constraints of external reality, and inner wishes displace outer perceptions. An example of this kind of error is the above-mentioned case where the "NO SMOKING" sign *became* a clock showing 5 p.m., because this accorded with the patient's wishes. His inner reality dominated over his external perception in a way that we do not normally allow. In the same way, his *wish* to meet his dead friend (or to be among friends) distorted his *perception* of a stranger in the hospital (someone whose features probably reminded him of his friend). Even when he recalled the fact of his friend's death, the external evidence could be put to one side in the service of maintaining the wish.

Primary process (mobility of cathexis) Primary-process "mobility of cathexis" can be defined as a situation in which feelings invested in one object are transferred to other objects without adequate constraint—usually in cases where the objects have some feature in common (sometimes quite a superficial feature). Such "mobility of cathexis" is apparent in the example where the patient conflates a stranger with his long-dead friend. Perhaps a better example comes from another patient, who clearly recognized her husband when he visited her in hospital, and treated him as such. Yet, when he was not there, she regularly referred to the man in the bed next to hers as being her husband, and behaved accordingly toward him. Again, the wish-fulfilling properties of such conflations are clear. She wanted her husband to be there. When he was, that was fine; but when he wasn't, it was not at all difficult to ignore or modify her conception of reality to fit with her requirements.

CONCLUDING REMARKS

These clinical phenomena reveal a number of important things about the mind and how it works. But, most important of all, they demonstrate the principle that it *is possible* to find the neurological correlates of some traditional psychoanalytic concepts and thereby to set them on a firm, organic foundation. The cases above show that the capacity of the "ego" to inhibit instinctual drives—the very foundation of rational, reality-constrained behavior—is somehow intimately bound up with the functions of the ventromesial frontal lobes. In subsequent chapters, we repeatedly pick up the strands of the issues raised by these cases and clarify further what they reveal about the functional organization of the mind. For now, having sketched a first, rough map of the neurological correlates of *consciousness* and *the unconscious*, we are in a position to consider in more detail, in the next chapter, what modern neuroscience reveals about "drives."

CHAPTER 4

EMOTION AND MOTIVATION

Our goal-directed actions are ultimately motivated by the biological task of meeting our needs in the outside world. The function of consciousness, described in the previous chapter, contributes a great deal to the successful performance of this task. "Core consciousness" relates information about the current state of the self to the prevailing circumstances in the outside world—the source of all the objects that the self requires to meet its inner needs. This information is *conscious* because it is *intrinsically evaluative*; it tells us how we *feel* about things. This applies especially to the inwardly derived aspect of consciousness—the conscious "state"—which provides our background sense of awareness. This background sense of awareness is not merely quantitative; it always has a particular qualitative "feel" to it. Conscious awareness is therefore grounded in *emotional* awareness.

WHAT IS EMOTION?

Emotion is akin to a sensory modality—an internally directed sensory modality that provides information about the current state of the bodily self, as opposed to the state of the object world. It adds a sixth sense (a sixth modality of "qualia") to our conscious existence. Emotion is the aspect of consciousness that is left if you remove all externally derived contents. If you were deprived of all sensory images (drawn from present and past perception), you would still be conscious. You would still be aware of your inner state—of your core *self*. Aristotle suggested that there are only five ways of knowing the world, corresponding to the five classical senses, but there is more to the world than the *outer* world.

EMOTION AS AN INTERNALLY DIRECTED PERCEPTUAL MODALITY

The "sense" of emotion is organized in a very different way from the externally directed sensory modalities. This is partly because it is a state-dependent function rather than a channel-dependent one. It reflects changes in your body that are communicated to the somatic monitoring structures of your brain, not only via discrete information-processing channels, but also via the gross chemical-transport mechanisms of the bloodstream and cerebrospinal-fluid circulation. These somatic monitoring structures, in turn, broadcast their outputs widely throughout the forebrain, thereby exerting a global "mass-action" effect on the information-processing channels of consciousness. (We have pointed out already, in chapter 3, that these outputs are not determined only by *actual* bodily events; body-mapping structures generate a *virtual* body, which is subject to all manner of chemical and other influences.)

Emotion is also different from the other sense modalities simply because it is *internally* directed. Only *you* can feel your emotions. This also applies to consciousness in general (cf. "the problem of other minds," chapter 2), but it applies to emotion in a special way. It is not only the *perception* of emotion that is subjective. *What* emotion perceives is subjective too. What you perceive when you feel an emotion is your *own subjective response* to an event—not the event itself. Emotion is a perception of *the state of the subject*, not of the object world. If a flash of lightning and clap of thunder cause you to feel a fright, it is not the lightning and thunder that you perceive emotionally (you *see* and *hear* them visually and aurally); it is your visceral *response* to those events that you feel emotionally. The same event can therefore make one person feel frightened and another not.

The fact that certain events make just about *everybody* feel roughly the same way is pregnant with significance for our understanding of the neurobiological mechanisms of emotion. We shall return to this fact in a moment.

MAPS OF THE BODY

The structures that form the *core* of the emotion-generating systems of the brain are identical to those that generate the background state of consciousness (see chapter 3). These phylogenetically ancient structures lie in deep regions of the brain, in the middle and upper zones of the brainstem (see Figure 4.1). The brain structures in question include the hypothalamus, ventral tegmental area, parabrachial nuclei, periaqueductal

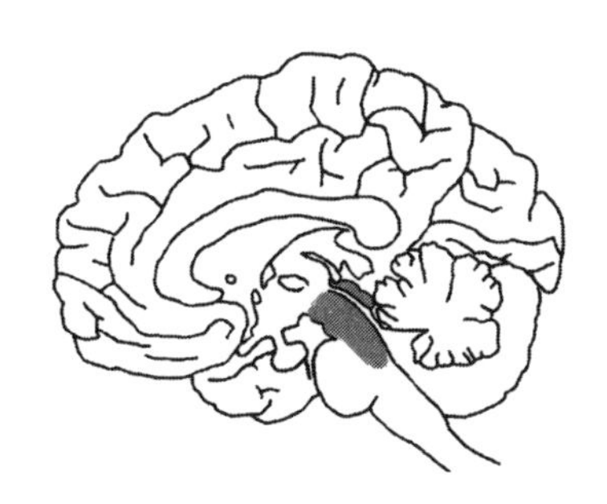

FIGURE 4.1
Location of core emotion-generating structures

gray, raphe nuclei, nucleus locus coeruleus complex, and classical reticular formation. As we discussed in chapter 3, all of these structures are implicated in the monitoring and regulation of visceral states.

Perhaps the most important of all these structures, as far as emotion is concerned, is the **periaqueductal gray** (PAG). This area of gray matter situated deep within the brainstem, surrounding the cerebral aqueduct (hence its name), has a vertical columnar organization (see Figure 3.2). The columns are divided into two broad types: some generate *pleasurable* sensations (in ventral [i.e. lower] PAG) and others generate *unpleasure* (dorsal [i.e. upper] PAG). Degrees of pleasure and unpleasure calibrate the basic qualitative range within which the "sense" of emotion is experienced. Pleasure and unpleasure might, therefore, be considered broadly equivalent to light and dark with respect to visual sensation, or high and low tones with respect to auditory sensation. It is important to note that *pain* (as we use the term) is not synonymous with *unpleasure*. "Unpleasure" denotes an *emotional* feeling (ultimately derived from the state of the internal milieu), whereas "pain" is a submodality of *somatic sensation*—one of the *externally* directed sensory modalities (see chapter 1).[1] Nevertheless, it is interesting to note that the PAG plays an important part in the generation of both unpleasure and somatosensory pain. This suggests that our exteroceptive consciousness of pain was, in an evolutionary sense, built upon the existing mechanism for generating unpleasure (or vice versa).

The distinction between unpleasure and pain reminds us of the fact that there are two sources of knowledge about the body,

[1]The fact that pain and unpleasure are not synonymous is perhaps best illustrated by pointing out that some people (sexual masochists) experience pain as *pleasurable*. Further evidence for this distinction is the fact that pain and unpleasure can be selectively targeted pharmacologically. Interactions between the somatosensory and emotional aspects of pain underpin the everyday travails of most pain clinics.

derived from its internal and external anatomy, respectively (see chapter 1). The first represents the "visceral" body—that is, the internal milieu. The internal milieu is regulated by various homeostatic mechanisms, which ensure that blood sugar, temperature, oxygen levels, and the like are adequately maintained. The state of these systems is what is monitored by the deep-brain structures listed above. These structures therefore generate a map of the *functions* of the body. The second source of bodily awareness is linked to the musculoskeletal system. This is the sensorimotor apparatus that moves the body around in the outside world. It is projected onto the cortical surface of the forebrain in much the same way as other objects in the external world are projected onto little maps of visual, auditory, etc., space. This generates a map of the *movements* (or potential movements) of the body.

These two sets of representations are not "maps" in a strict topographical sense—that is, they are not scale models of bodily anatomy.[2] The map of the internal milieu, in particular, does not represent the body topographically at all; it gathers together and represents pertinent information about the body's homeostatic *physiology*, not its musculoskeletal *anatomy*.

---

[2] On a map of Britain, the ratio of distances from London to Cambridge to Edinburgh is the same as occurs in reality—that is, the map retains an accurate record of the topographical relationship between these spatial elements. However, not all maps work like this. The map of the London Underground system (perhaps the most famous nontopographical map in the world) does not provide spatially accurate topographical information. The ratio of distances between Paddington, Baker Street, and King's Cross stations on the map is not the same as that found in reality—though obviously some spatial properties, such as the sequence of these stations on a line, is retained. Although the metric properties of the map have been distorted (to ensure ease of use), no one would claim that the map of the London Underground is not a "map." It contains relevant information about objects of importance and organizes the information in a way that allows it to be usefully employed.

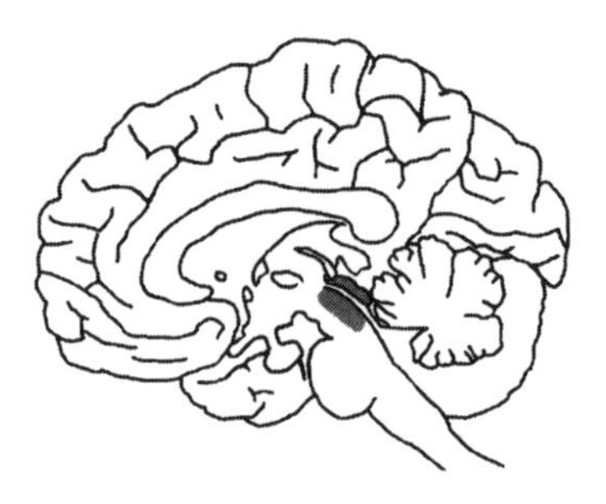

FIGURE 4.2
Tectum and dorsal tegmentum

We have already listed the structures that map the body's physiological functions. Maps of the body in the second (topographic) sense are found in several locations in the brain. One of them is of special interest to us, for reasons that will soon become apparent. This map is located in the **tectum** and **dorsal tegmentum** of the upper brainstem (Figure 4.2)—that is, close behind the *ventral* tegmental area depicted in Figure 4.1. This region of the brain receives input from *all* the sensorimotor modalities, and it is therefore one of the "zones of convergence" mentioned in chapter 2 in connection with the "binding problem." The proximity of this map of the musculoskeletal body to the adjacent projections of its visceral states is important, for two reasons.

First, these two maps together generate a rudimentary representation of the *whole* person, the inner and outer "virtual bodies" combined. Jaak Panksepp—a leading neurobiologist whose conceptualization of the functional anatomy of emotion is very similar to that of Antonio Damasio—therefore (rather boldly) calls this brain region the **SELF** ("*S*imple *E*go-like *L*ife *F*orm": Panksepp, 1998). (He capitalizes the term to distinguish the technical from the colloquial meaning of the word.) This primal SELF forms the foundational "ego" upon which all our more complex representations of our selves are built (cf. the "autobiographical self" described in chapter 3).

The second reason why the proximity of the two body maps is important is that it gives the emotion-generating part of the brain direct access to one of its *action*-generating mechanisms. The homunculus in the dorsal tegmentum provides a combined sensori*motor* map of the body, which generates primitive action tendencies (e.g., approach and avoidance behaviors, which are closely linked to pleasure and unpleasure, respectively). This reminds us of the very important fact that we do not only *experience* our emotions, we also *express* them. Emotion is not only an internally directed perceptual modality, it is also a form of motor discharge.

EXPRESSION OF THE EMOTIONS

The perceptual aspect of emotion has a compulsive effect on us. We cannot simply lie back and feel our emotions. They make us want to *do* something. This "doing something" once again implicates the inner/outer distinction to which we have referred so often. The motor aspect of emotion involves both inwardly and outwardly directed discharge processes. Inwardly, the experience of emotion is accompanied by the release of hormones, changes in breathing and heart rate, vasodilation and vasoconstriction, changes in regional blood supply, and so forth. Outwardly, emotion manifests itself in various ways: through changes in facial expression, baring of teeth, crying, blushing, and the like, but also in complex behaviors like shouting, running away, and lashing out. Some of the inwardly and outwardly directed discharges of emotion are difficult to distinguish from each other (e.g., laughing, crying, blushing). It is also difficult to draw a sharp distinction between the perceptual and motor aspects of emotion; for example, the feeling of a racing pulse is an important part of the *perceptual* complex of fear, as is the urge to run or hide.

We mentioned before that the fact that certain events make just about *everybody* feel roughly the same way is pregnant with significance. This applies equally to the perceptual *and* motor aspects of emotion. Certain situations are apt to evoke certain feelings in all of us, and they are equally apt to make us want to act in certain relatively stereotyped ways. For example, the sight of a snake slithering rapidly toward you is likely to evoke a feeling of fear, no matter who you are, and it also makes you freeze all bodily movement. Such situations appear to have universal significance. Our ability to recognize them, and our reactions to them, appear to be largely innate.[3]

Neurobiologists call these universal affective reactions **basic emotions**. The "basic emotions" appear to consist in "hard-wired" *connections* between certain external situations of biological significance and the subjective responses they evoke. This implies that certain patterns of external perceptual stimuli are innately linked to specific internal perceptual stimuli, and that these perceptual links automatically trigger innate (both internal and external) motor responses. The orchestration of these different elements of the "basic emotions" unfolds over concrete anatomical pathways and involves specific physiological mechanisms. Bearing in mind what we have said already about the anatomy and physiology of emotion, the reader can perhaps predict which structures are likely to be involved. For example, we can safely assume that the orchestration of outer and inner perceptions will involve links between extroceptive forebrain and introceptive brainstem structures, and that the PAG will be centrally involved in the generation of most (if not all) the basic emotions.

By virtue of these fixed anatomical arrangements, the basic emotions have provided a powerful research paradigm for understanding the brain mechanisms of our inner mental life.

---

[3]This does not mean that these inborn emotional action tendencies are not *modifiable* (see below).

EMOTION IN MAN AND IN OTHER ANIMALS

Needless to say, the "basic-emotion command systems" (as they are called) evolved over eons of time. The basic emotions exist because they have established survival value. In situations of biological significance (e.g., mortal danger, proximity of a fertile mate), these emotions provide ways of reacting that increase the likelihood that the organism will survive and reproduce, and thereby propagate its genes. For this reason, Panksepp (1998) suggests that the basic emotions should be thought of as "e-motions"—"evolutionary motions." Precisely because these mechanisms take so long to develop, and also because they have such potent survival value, they are deeply conserved in mammalian genotypes. They surely evolved long before *Homo sapiens* even appeared on the evolutionary tree, and they will long continue to be preserved.

Accordingly, we share with all other mammals the basic-emotion command systems (and the feelings that correspond to them) described in the following sections of this chapter. Dogs, cats, dolphins, whales, rats, mice—all possess the anatomical and physiological arrangements we describe in the following sections. This shared evolutionary heritage literally embodies the primal experiences of our ancestors, which, even if we cannot reexperience them, have left traces in our "procedural-memory" systems (see chapter 5).[4] The basic emotions therefore define a

[4] Freud's belief in phylogenetic memories (the "inheritance of acquired characteristics") has often been derided. He seems to have believed (incorrectly) that it was the *frequency of occurrence* over generations of these primeval events, rather than the survival value of certain ways of responding to them, that led to them being preserved. The figurative language that Freud (1912–13) used to describe this aspect of mental life also created the (false) impression that episodic memories of the events themselves were literally "passed down." The influence of such primeval events on our procedural or instinctual memory systems is certainly a reality, but the indirect mode of transmission and influence does not permit literal "remembering" (see chapter 5).

set of common biological "values" that unite us all in our struggle with the tasks of life. (Cf. the question of consciousness in other animals, in chapter 3.)

THE BASIC EMOTIONS

The scientific data upon which this knowledge is based comes from observations of anatomical structures that produce reliable emotional effects when *modified* in some way. The data come from both animal and human neuroscience, as well as from biological psychiatry.

In animal research, the "modification" might consist of increasing the activity of a structure by stimulating it electrically, or administering quantities of the chemical messenger that is typically employed to excite a system. Neurological results are recorded and linked with observations of the animal's behavior. (It is impossible to monitor an animal's subjective state.) The activation of a structure can also be *reduced* by administering a chemical that blocks its normal activity, or a structure can be eliminated by removing it surgically or ablating it chemically.

When the same brain regions are investigated in humans, the results are typically highly consistent with the animal work. Research with humans lacks the precision of animal research because human researchers do not selectively ablate circumscribed areas of brain tissue. We must investigate people in whom natural events (like strokes or tumors) have produced similar effects. These lesions are usually not circumscribed to a single structure. Likewise, chemical manipulation of the human brain, such as can be observed in drug users and psychiatric patients, usually lacks the specificity and selectivity that is achieved when animal researchers deal with brain tissue more directly. In the case of humans, though, we have access to verbal reports of the *subjective states* that occur when the emotional parts of the brain are modified. There are also a good few studies of the subjective effects of local brain stimulation during surgical operations and in cases of focal epilepsy.

Readers who are skeptical of the idea that one can simply "turn on" or "turn off" an emotion are encouraged to take a closer look at this literature.[5] This field of research has produced a far better understanding of the neurobiology of emotion than we could ever have hoped for. Although there are still controversies in the field, especially on the borderlands, the findings reported below are not really controversial; they represent the conservative "basics" that most neuroscientists working in this area would now agree upon.

There appear to be four "basic-emotion command systems" in the brain. In the following sections, we use the nomenclature of Panksepp (1998) to describe these systems: SEEKING, RAGE, FEAR, and PANIC. Some of the terms that other authorities use for these emotion systems are also mentioned. By deriving a common denominator from these alternative terms, the reader should get a good idea of each of the emotions we are talking about.

The SEEKING system

Long known as a "reward" system, the SEEKING system is also associated with the terms "curiosity," "interest," and "expectancy." This system provides the arousal and energy that activates our interest in the world around us. On the perceptual side, it generates the feeling that something "good" will happen if we explore the environment or interact with objects. On the motor side, it promotes exploratory behaviors, like foraging.

[5] Panksepp's (1998) masterful survey of the field is perhaps the best place to start. Joseph LeDoux's highly readable 1996 book *The Emotional Brain* is perhaps more accessible, but it focuses more narrowly on a single emotion: fear.

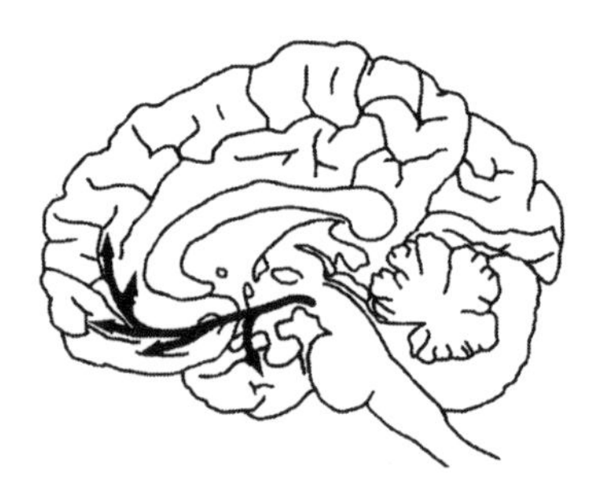

FIGURE 4.3
The SEEKING system

Exploratory behavior varies somewhat from species to species, and it also depends on the particular need that is activating the system at a given moment, but it typically involves sniffing, touching, and oral exploration. This system is heavily activated during sexual arousal and other **appetitive** states (e.g., hunger, thirst, or even craving a cigarette).[6] It is also associated with play, especially of the rough-and-tumble variety, and some forms of aggression (especially the predatory variety, known as "cold" aggression).

The source cells of the SEEKING system are located in the ventral tegmental area (Figure 4.3). The axons of these cells pass through (i.e., synapse on) the dorsolateral hypothalamus, *en route* to the nucleus accumbens, which is where most of these axons terminate. They also project further upward to the anterior cingulate gyrus and other cortical areas in the frontal lobes, and downward to the amygdala (in the temporal lobe).

The command neurotransmitter of this system is *dopamine*. (The SEEKING system forms part of the *mesocortical–mesolimbic DA system* mentioned in chapter 1.)

[6]Where Freud used the sexual term "libido" to denote the mental function activated by our bodily needs of all kinds, modern neurobiologists speak of "appetites."

The neurobiology of "libidinal drive" The concept of **drive** seems to be unfashionable in psychoanalysis nowadays. It is unclear why this happened, but it has had the unfortunate result of divorcing psychoanalytic understanding of the human mind from knowledge derived from all other animals. We humans are not exempt from the evolutionary biological forces that shaped other creatures. It is therefore difficult to form an accurate picture of how the human mental apparatus really works without using a concept at least something like Freud's definition of "drive":[7]

> The psychical representative of the stimuli originating from within the organism and reaching the mind, as a measure of the demand made upon the mind for work in consequence of its connection with the body. [Freud, 1915c, p. 122]

This definition conveys very well the place that the SEEKING system occupies in our mental economy as a whole.

How is the SEEKING system normally activated? There is a range of **need-detector** mechanisms in the hypothalamus (particularly the lateral and ventromedial nuclei, with extensive connections to other brain regions). These detectors constantly "sample" the internal milieu to maintain its delicate economy. Different hypothalamic regions switch these detector systems on (they act like "accelerators") and off (acting like "brakes"). One of these systems, for example, regulates your core body temperature. It ensures that your temperature stays in the (very narrow) correct range. There is also a thirst detector, a hunger detector, even a "sexual-need" detector. To illustrate how these systems work: Lesions of the "brake" hunger system produce a dramatic

[7] Freud used the German term *Trieb* where we say "drive," but his English translators preferred to use "instinct." Where Freud spoke of "libidinal drives," we nowadays use the tem "appetitive drives." Appetitive drives activate the SEEKING system, discussed here.

increase in an animal's interest in food. Given unlimited access to supplies, it will eat to the near total exclusion of interest in all other events in the world. Gross obesity follows very rapidly, at which time food consumption slows. Conversely, lesions to the "accelerator" system create an almost total loss of interest in all food. Anorexia follows rapidly, although the animal will occasionally nibble—just enough to remain alive.

It is uncertain how specific each detector system is to its target need—for example, whether the "thirst detector" is concerned only with thirst. They are probably not entirely specific, but the important issue is that these hypothalamic systems generate "needs," and these "needs" activate the SEEKING system. In other words, when the need-detector systems register that one of the homeostatic mechanisms they monitor has moved out of its acceptable range, they activate seeking—"appetitive"—behavior to correct it. This activation can be maintained for long periods of time. The SEEKING system can also be activated by a range of other inputs, both perceptual and cognitive, but the simple "detector" mechanisms provide the best illustration of how this system works.

What does the SEEKING system do? As the name suggests, it seeks. The more difficult question is: *What* does it seek? One might think that it seeks the specific object of a current need, as determined by the need detectors. The reality is slightly more complex. The SEEKING system itself does not appear to know what it is seeking. (In psychoanalytic parlance, one might say that it is "objectless.") The SEEKING system appears to be switched on in the same way by *all* triggers, and, when activated, it merely looks for *something* in a nonspecific way. All that it seems to know is that the "something" it wants is "out there." A nonspecific system like this cannot by itself meet the needs of an animal. It has to interact with other systems. The mode of operation of the SEEKING system is therefore incomprehensible without reference to the *memory* systems with which it is intimately connected. These systems provide the *representations* of objects (and past interactions between the self and those objects) that enable the organism to *learn* from experience. One of the most basic tasks that these combined systems have to perform is to distinguish which objects in the outside world possess the specific properties that the internal milieu lacks when a particular need detector switches "on." Like any system of learning, this requires a "reward" mechanism. Panksepp labels this extension of the SEEKING system the LUST system.

The LUST subsystem The LUST subsystem has a longer history of being called a "pleasure," "reward," or "reinforcement" system. These terms reveal that the function of the system is associated with *gratification*—that is, with *consummation* of the appetites that activate the SEEKING system. On the perceptual side, this system generates feelings of pleasurable delight: "That feels *good*!" On the motor side, this system switches appetitive behaviors off and replaces them with **consummatory** behaviors. (There is a reciprocal relationship between SEEKING and LUST activation.) Like exploratory behaviors, consummatory behaviors are complex reflexive-action programs that vary somewhat across species (and the sexes) and across needs. These instinctual behavior patterns are automatically released when the object of a biological need is attained. The thirsty cat laps up milk; the sexually aroused male dog rhythmically thrusts its penis.

The LUST system consists of a complex group of structures arising from the hypothalamus and lying mostly in the basal forebrain, close to the main termination of the SEEKING system's ascending projections (see Figure 4.4). The most critical of these structures appear to be parts of the septal region and hypothalamic nuclei (mainly the preoptic area). Stimulation of these structures (in humans) produces orgasmic feelings. The system terminates in the PAG, which is presumably where the

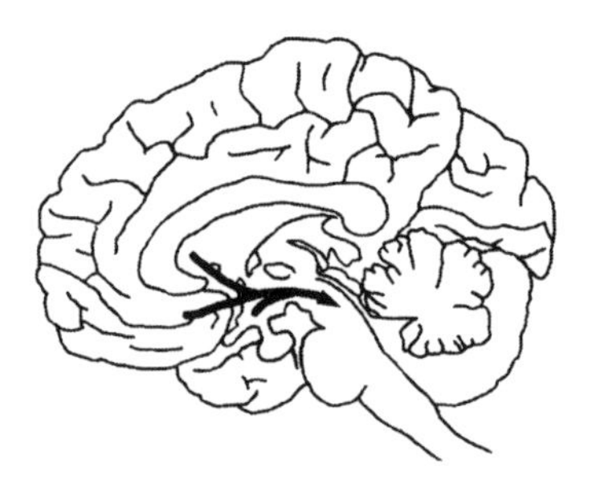

FIGURE 4.4
The LUST system

pleasurable sensations are in fact generated or "perceived" (i.e., where the pleasure centers exert their influence on the virtual body of the primal SELF). The command neuromodulator (actually, neuropeptide) of this system is **endorphin**.

As with the whole concept of "basic-emotion command systems," the idea of "pleasure centers" in the brain might seem overly simplistic—even unlikely—to psychoanalytically minded readers who are familiar with the complexities of human desire. It is therefore important to point out that these crude mechanisms (the existence of which is beyond doubt, and which we share with all other mammals) are subject to a wide range of higher cognitive influences that can modulate, modify, and inhibit them and their associated behaviors in multifarious ways (see below).

The key finding that revealed the existence of these pleasure centers dates back to the 1950s and the work of James Olds (Olds & Milner, 1954). In a series of (initially serendipitous) experiments, Olds demonstrated that animals are prepared to work extremely hard in order to experience electrical stimulation of these brain regions, especially if the electrodes are placed in the septal nuclei. This experimental behavior is called **self-stimulation**. Animals rapidly learn the skilled movements necessary to switch on a neurostimulator when the electrodes are placed in these regions, and they continue to stimulate themselves to exhaustion, and to the near total exclusion of any other behavior. Even when offered a choice between behavior (usually bar-pressing) that produces either food, water, sex, or self-stimulation, they overwhelmingly choose self-stimulation. Typically, such animals self-stimulate for long periods, then occasionally shift to the bar that offers food for a brief snack before quickly returning to self-stimulation. Needless to say, these animals rapidly lose a great deal of weight. The parallel with *addictive* behavior in humans is obvious.

Drug addiction The SEEKING and LUST systems are designed to promote learning, and they motivate us to acquire the skills that are necessary to meet our inner needs in the outside world. The need detectors of the hypothalamus activate the SEEKING system so that this system might generate behaviors that are apt *to meet our actual biological needs* (nutritional needs, the need to reproduce, etc.). Likewise, the pleasure centers of the basal forebrain transmit pleasurable sensations to signal to the brainstem SELF that the object required to meet its actual bodily needs has been attained. The "rewarding" quality of these sensations also *motivates* the animal to perform the *work* necessary to attain these biological objectives. The generation of pleasure *for its own sake* serves no biological purpose. Just like self-stimulation behavior in experimental animals, the use of recreational drugs—such as **cocaine** and **amphetamines** (which stimulate the SEEKING system and thereby artificially generate positive expectancies) and *heroin* and other **opiates** (which stimulate the pleasure centers *directly*)—hijacks, or short-circuits, these adaptive mechanisms. These drugs generate pseudoappetitive behaviors (and associated cravings) and pseudoconsummatory behaviors (and associated pleasurable sensations) that serve *no biologically useful purpose.*[8]

Other psychopathologies The SEEKING system of a newborn baby is switched on when activated by a need, without the baby knowing *what* is needed. Left to its own devices, it is so helpless that it will never find the objects required to satisfy its needs and will therefore die. For this reason, it has caregivers that function as "intermediaries" between its needs (communicated by the expression of its emotions) and the objects in the outside world. The actions that these intermediaries perform on the baby's behalf—and their effects—are then gradually learned ("internalized") until the child can take care of itself. This, as we all know, is why parenting is so important. Early experiences of satisfaction form the templates of our understanding of how life works; for a child, learning how to adequately recognize its needs and meet them in the world is utterly bound up with the quality of the parenting it receives. There are all sorts of subtle ways in which this process might be disrupted or distorted (for instance, if a baby's needs are routinely neglected or misunderstood or even met too soon, before they can be felt). The foundations can thereby be laid for later psychopathology—in combination with a

[8] Some of the dangers of recreational drug use arise from the fact that (as with self-stimulation behavior in animals) preoccupation with the drug can become nearly all-consuming, to the exclusion of all other (biologically useful) behaviors. Also, just as laboratory animals are prepared to work extremely hard to activate the pleasure system, addicts will go to great lengths to obtain another "fix"—as demonstrated by the fact that they resort to burglary and prostitution. Other dangers arise from the fact that the SEEKING and LUST systems can become tolerant of these drugs, with the result that greater use is required to produce the same effects. The drugs in question can also have other, more directly damaging effects on the brain and other bodily tissues (e.g., toxic effects).

set of biological "risk factors," such as variation in the inherent "setting" levels of the basic affective systems.

Even in states of rest, the SEEKING system is tonically active to a certain extent—as long as you are alive, you always need something. Sustained underarousal of the SEEKING system is associated with pathological lack (or loss) of interest in the world. Likewise, unmodulated hyperarousal of this system can produce overexcited states or excessive interest in inappropriate objects and activities. Whatever their cause may be, disregulations of this type can be managed psychopharmacologically. Drugs that modify mesocortical–mesolimbic dopamine transmission, in various complex ways, therefore form the basis for entire classes of psychiatric medication, used in the treatment of not only the schizophrenias, but also attention deficit and hyperactivity disorders (ADHD), tic disorders, and mood disorders, underlining the vital role that this system plays in human motivation and emotion. (Some of these points will be returned to in our discussion of dreams and hallucinations in chapter 6.)

Activation of the LUST system switches the SEEKING system "off," signaling that an inner need has been met. Activation of the other basic-emotion command systems occurs when our drives are *unmet*, in various ways. Accordingly, the activation of these other systems is associated with varieties of *unpleasure*. The particular variety of unpleasure (and associated instinctual discharge) that is released depends on the nature of the biologically undesirable experience that activated it.

The RAGE system

More than any other system, the RAGE (or "anger-rage") system is activated by states of *frustration*—when goal-directed actions are *thwarted*. The term "anger-rage" is used to denote the *feeling state* associated with the arousal of this system. This term is necessary because not all aggressive behavior is activated by the RAGE system. Neurobiologists distinguish between two (or three) different types of **aggression**. The RAGE system is associated with only one of them: so-called "hot" aggression. The "cold" type of aggression, associated mainly with predatory behavior, has little to do with feelings of anger or rage; rather, it has to do with appetitive *seeking* and is therefore driven by the dopaminergic system described above. (There is a third variety of aggression, associated with *male dominance behavior*. Neurobiologists classify this type of aggression with the "social emotions," some of which will be discussed later.) The fact that aggression has at least two different neural substrates must have some important implications for psychopathology (for forensic psychology and psychiatry). Here, therefore, is another fruitful area for future collaborative research (see chapter 10).

Feelings of anger-rage (the perceptual aspect of this system) release stereotyped motor programs associated with the well-known "fight" (as opposed to "flight") response. The fighting response is also called "affective attack" behavior. Externally, this involves a facial grimace with baring of teeth, commonly accompanied by an aggressive-sounding noise (e.g., a growl). The body adopts a stable, broad-based posture, and the claws (or fists!) are extended. Internally, there is a series of modifications in the autonomic nervous system—such as increased heart rate and redirection of blood supply to the skeletal musculature needed for violent "action" situations (at the expense of the "action-irrelevant" digestive system)—which allow the animal better to engage its enemy.

These changes are orchestrated by amygdalar projections to the PAG. As stated above, the amygdala (in the temporal lobe) is one of the terminal projections of the SEEKING system, and it is made up of a number of different nuclei. The key structure involved in triggering anger-rage is the *medial* nucleus of the amygdaloid complex. This system courses through the bed nu-

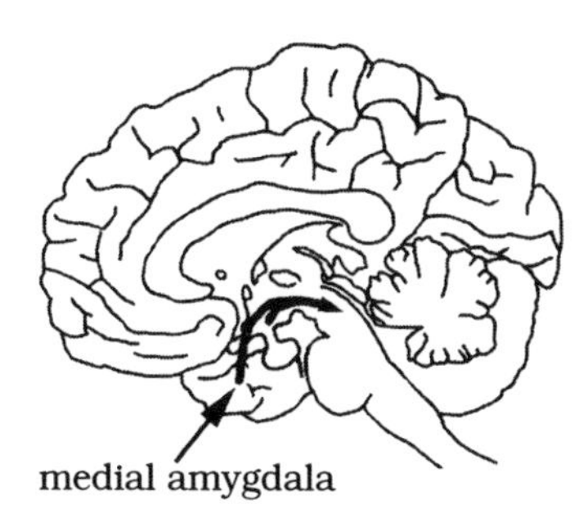

FIGURE 4.5
The RAGE system

cleus of the stria terminalis and (anterior, ventromedial, and perifornical) hypothalamus before it, like all the other basic-emotion command systems, projects down into the (dorsal) PAG (see Figure 4.5).

Unlike the SEEKING system, this system is activated only sporadically. However, when it *is* tonically activated at a low level, we use the word "irritability" to describe its effects. Like anger-rage, irritability is usually caused by the frustration of goal-directed activities. The person experiences background annoyance, and the residue of the motor output of the system manifests in behaviors like scowling and/or tensing of the muscles—especially the hands. This form of chronic low-level activation of the anger-rage system, which **primes** it for full-blown affective attack, seems to be a frequent accompaniment of modern life (perhaps especially in large cities).

It is not difficult to see the evolutionary advantages of having a system like this "hard-wired" into the brain. Instead of each generation having to learn afresh the mechanics of how best to deal with an aggressor, this neurobiological cicuit preprograms an entire set of automatic "output" routines. The animal possessing such a system is far more likely to survive its first violent encounter than the one that has to learn all these routines from scratch. The programs that such systems encode were selected and preserved through evolution because they represent multi-purpose perception and action systems that are applicable to a wide range of typical mammalian life-events—from competing with peers for food and sexual partners to avoiding being eaten by predators.

The FEAR system

The second *negative* emotion command system is probably the most intensively researched system of all (for a review see LeDoux, 1996). It generates (on the perceptual side) feelings of fear-anxiety and (on the motor side) the "flight" response. Just as the different substrates of "hot" and "cold" aggression taught us to distinguish between different forms of violence, neuroscientists have also learned to distinguish between *fear-anxiety* and *panic-anxiety*. (To a certain extent, these two varieties of anxiety correspond to the psychoanalytic distinction between "paranoid" and "depressive" anxiety.) The **benzodiazepines** (minor tranquilizers, like diazepam) are successful in reducing fear-anxiety—by increasing GABA inhibition at certain receptors. Panic-anxiety, by contrast, responds mainly to *antidepressant* medications.

Like the RAGE system, the FEAR system is centered in the amygdala and its connections (see Figure 4.6). The *lateral* and the *central* nuclei of the amygdaloid complex are the hub of this system. (The balance between "fight" versus "flight" responses is apparently determined by interactions between the lateral-central and medial parts of the amygdala.) From there, the circuit projects through the (medial and anterior) hypothalamus before it terminates in the (dorsal) PAG of the brainstem—which is where the feelings in question are actually generated ("perceived" by the SELF) and the motor programs released.

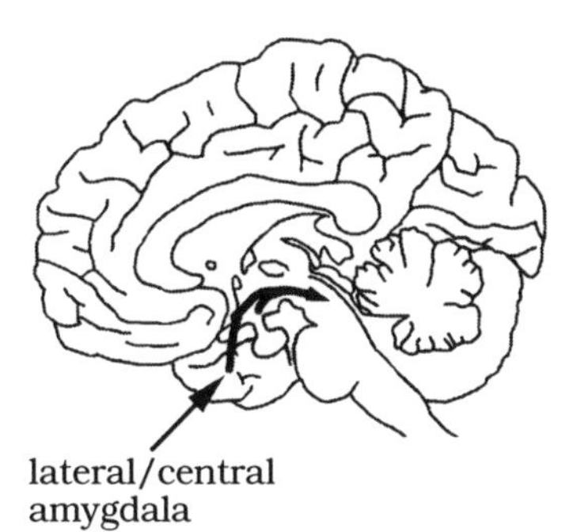

FIGURE 4.6
The FEAR system

The evolutionary advantages of this system are that it allows us to escape rapidly from dangerous situations and to avoid such situations in the future. On the perceptual side, brain stimulation of the sites listed above (in humans) is associated with feelings of extreme anxiety or terror. On the motor side, stimulation of these sites typically makes an animal run away and try to hide. Mild stimulation, on the other hand, leads to a "freeze" response; this is probably because many predators recognize objects through motion, and assuming a motionless posture can therefore help to avoid detection. "Freezing" is, however, only a good strategy under some circumstances. When a predator is quite far away, immobility is generally a useful strategy; however, if an approaching predator is nearby, there is little point in remaining motionless, and flight is the better option. These are the external motor manifestations of fear-anxiety. Internally, there are a range of changes in the viscera, which mirror those seen with activation of the RAGE system: increased heart rate, more shallow and rapid breathing, and redistribution of blood from the gut to the skeletal muscles. The latter change can cause diarrhea in cases of extreme fear, especially when the affect is sustained over a sufficiently long period.

Fearlessness The functional anatomy of the RAGE and FEAR systems allows us to understand some fascinating findings from patients with neurological disorders. A few patients have been described who have selective bilateral lesions of the amygdala, which is usually a result of a rare metabolic condition. These patients are normal individuals in almost all respects. They are, however, distinguished from the norm by the fact that they lack anger-rage and fear-anxiety—the two "negative" emotions mediated by the amygdala. One such case is described by Adolphs, Tranel, and Damasio (1994; see also Damasio, 1999b, pp. 62–67). Their findings—which focus mainly on fear-anxiety—are remarkable. The patient is a bright woman, who understands very well, at a cognitive level, what the concept of "fear" is. At a perceptual-motor level, however, she cannot recognize fear in the faces of others, and she cannot generate a fearful facial expression herself—not even when the examiner models it for her. Her performance in recognizing and producing facial affect is normal for all the other commonly understood emotions, like happiness and sadness. More importantly, her *behavior* is totally devoid of fear. She is exceptionally friendly and is far more prone to touching and hugging than is common among other patients visiting the authors' laboratory. She is willing to interact with virtually anyone who engages her in conversation and finds it easy to trust people she has only just met. Unfortunately, but not surprisingly, she has frequently had her trust in others abused.

In more severe cases of this condition, which is known as the **Kluver-Bucy syndrome**, in addition to displaying fearless and angerless placidity, such patients (like amygdalectomized laboratory animals) become *hypersexual*. There is a dramatic increase in the *amount* and *variety* of their sexual behavior—so that objects that would not previously have been attractive to them (e.g., members of the same sex, other species, even non-living objects) now are—and there can also be a dramatic increase in masturbation, including in public. These patients (and animals) also become *hyperoral.* They indiscriminately explore objects by mouth and sometimes attempt to eat inedible objects. In addition, they display a symptom called *hypermetamorphosis*—which means that they become hyperdistractible, as everything seems to be of equal interest to them. In animals (but not in humans, due to differences in the anatomy of vision), the symptom of *visual agnosia* (inability to recognize objects visually) is an associated feature of this syndrome.

The personalities of these patients are obviously and dramatically skewed by their neurological conditions. They demonstrate how critically important for normal mental life the "negative" functions of the RAGE and FEAR systems are. This sort of evidence helps us to establish the neural correlates of certain aspects of personality (see Kaplan-Solms & Solms, 2000) and may also help us better to understand the ways in which genetic and environmental factors modify the biological systems that control personality.

The PANIC system

The PANIC (or *separation-distress*) system is associated not only with panic-anxiety, but also with feelings of loss and sorrow. This provides neuroscientific evidence for the link that psychoanalysts have long recognized between panic attacks, separation anxiety, and depressive affect. The operation of this system seems to be intimately connected with *social bonding* and with the process of *parenting*—for reasons associated with the neurochemistry of the system and the way it is designed to operate.

The core of the separation-distress system is the *anterior cingulate gyrus*, which has extensive connections with several thalamic, hypothalamic, and other nuclei (see Figure 4.7)—including the bed nucleus of the stria terminalis, preoptic hypo-

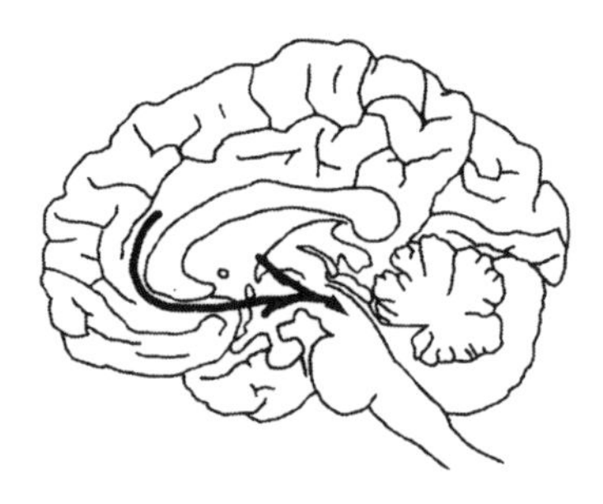

FIGURE 4.7
The PANIC system

thalamus, and ventral tegmental area. These sites are known to be of importance for sexual and maternal behavior in lower mammals. As in all the basic-emotion command systems, there are onward connections from these regions to the (ventral) PAG. The neurochemistry of this system is dominated by the endogenous **opioids**. There is also some evidence that **oxytocin** and **prolactin** are centrally involved in the operation of the system, which, as we shall see in the next section, further underlines the link between the PANIC system and maternal behavior.

Stimulation of some of these structures (in humans) has been associated with sudden onset of panic attacks, and even—in one case—a full-blown clinical depression, which met all the DSM-IV criteria. The depression recovered completely with offset of the stimulation. In animals, stimulation of this system produces "distress vocalizations," or "separation calls." These vary from species to species, but they involve actions such as crying, howling, or chirping. Sustained stimulation of this system in freely roaming animals produces an interesting sequence of behaviors. Initially, when the system is first stimulated, it promotes SEEKING behaviors, together with distress vocalizations. Presumably this increases the chances of finding the mother, or of being found. After a more or less set period of time, there is a change to *withdrawal* from the environment as the animal retreats into isolation and displays a sort of hibernation behavior that looks for all the world like depression. This shift in the animal's behavior, from seeking to withdrawal, is presumably related to the fact that it is not safe to be looking for your mother for too long, due to the risk of attracting the attention of a predator. If she is not around, the best thing to do is to lay low and wait for *her* to find *you*.

The role played by endogenous opioids in this system teaches us an interesting lesson about the nature of attachment. This is best conveyed by describing the various ways in which reinforcement can operate—matters long investigated by animal behaviorists. It is easy to teach animals to do something if they are rewarded when they perform an experimental task. In learning theory parlance, this is called "positive" reinforcement. There is another type of reinforcement that is also a very effective route to learning. This is called "negative" reinforcement. Here, the animal receives *constant* reinforcement, and inappropriate behaviors are associated with the *withdrawal* of this reinforcement.[9] It is this type of schedule that appears to govern the separation-distress system. Endogenous opioids (like externally administered **morphine**) reduce feelings of pain. Separation from a loved object increases activation of the PANIC system, which reduces the level of opioid activity in this part of the brain. Separation and loss, therefore, are literally "painful." Young animals rapidly learn to maintain proximity to their caregivers by virtue of this reinforcement strategy.

The CARE subsystem This "social" emotion system influences the *mother's* behavior no less than the *child's*. We know that levels of oxytocin and prolactin, two of the key chemicals that

[9] This is not the same as *punishment*—where an action is linked with a bad consequence. As many parents know, punishment is not an effective method of promoting learning.

govern this system, are increased in the period around birth. This increases the extent of mother–infant *bonding* during the immediate postpartum period—again, an aspect of the operation of this system with clear evolutionary advantages. Of particular interest is the fact that these same chemicals are strongly implicated in female *sexual* behavior. This underscores the sexual underpinnings of mother–infant intimacy, which has long been of interest to psychoanalysis.

There are also interesting clinical implications relating to modifications of this system in some children. This opioid system has been found to be *overactive* in some cases of autism. Consequently, such children experience far less "pain" on separation than their peers, and as a result they bond less well with caregivers and other people. Consistent with this, drugs that block the operation of opiate channels produce more positive social interactions in some cases of autism. But, importantly, the drug only appears to work (to the extent that it can) if it is combined with renewed, facilitating encouragement from the social environment. It is as if the drug opens a window, but by itself it cannot change the nature of the child's object relationships (Panksepp, 1998).

PLAY and other social emotions

As the previous section suggests, neuroscientists are beginning to extend the "basic-emotion" paradigm into other, more complex aspects of human ethology. Perhaps the most interesting avenue of current research in this direction concerns what Panksepp (1998) calls the PLAY system. It is a remarkable fact that all young mammals (including humans) seem to *need* to play, and to need a certain *amount* of play. Whatever its biological purpose may be, play (and especially rough-and-tumble play) seems to function in young children according to homeostatic principles similar to those that regulate such basic functions as sleep. If a rat pup is deprived of opportunities for rough-and-tumble play, this is followed by a *rebound* effect whereby the pup makes up for the lost time by playing proportionately more whenever it is next given the chance. The fact that this mechanism is so deeply conserved in the mammalian series suggests that play probably performs some critically important developmental functions. Panksepp suggests that the veritable epidemic of ADHD in modern American cities may in part be a consequence of these children being deprived of adequate amounts of rough-and-tumble play.

LEARNING FROM EXPERIENCE

In the foregoing sections, we have repeatedly mentioned the evolutionary advantages of these inherited, emotionally driven behavioral stereotypes. What needs to be emphasized, however, is that it is not enough to have only four emotional responses—SEEKING, RAGE, FEAR, PANIC—coupled with a handful of automatic, stereotyped behaviors to cope with the vast complexities of everyday mammalian life. The world is almost infinitely unpredictable, and we must modulate and regulate ourselves accordingly.

This is reflected in the fact that all the basic-emotion command systems discussed above are, to variable degrees in different species, but to a very great degree in humans, open to influence by *learning* mechanisms. In other words, although these systems are innate, they are by no means "hard-wired" in the sense of being *unmodifiable*. On the contrary, they appear to be specifically designed in a way that requires "blanks" to be filled in by life experience (and especially early experience). This general topic is discussed in detail in chapters 5 and 7, so only a few specific points are made here.

We have already illustrated the essential points we need to make when we described the role of learning mechanisms in relation to the "objectless" drives of the SEEKING system. The young animal knows *that* it needs but not *what* it needs—it has to learn from experience which objects in the world satisfy its needs and which do not. The evolutionary advantage of this is that it enables the animal to adapt to whatever environment it is born into, where the types of available satisfying object may vary widely. Youngsters (and especially those with protracted periods of infantile motor helplessness, like humans) are unlikely to survive this early learning process without the mediation of adult caregivers, who actively *teach* the little one what it has to do to meet its inner needs and survive the attendant dangers. We have also mentioned how easily this mediation process can go awry, and how devastating the consequences of this might be for the future mental health of the child.

Similar considerations apply to the other basic-emotion command systems. In the FEAR system, for example, although some dangerous objects and situations *do* appear to be hard-wired into the system (hence the stereotyped nature of most phobias), the representational (or "object") aspect of the system is left largely blank, to be filled in by early experience. LeDoux (1996) describes in detail how this process works. Two points are of special interest.

The first point is that the connections that link the noxious stimulus (the object-to-be-feared) with the fear-anxiety responses are made with *extreme rapidity*, and they are thereafter maintained *outside extended consciousness*. Once a stimulus (thing or place) is associated with a painful experience (sometimes on the basis of just *one* exposure to the noxious stimulus), the FEAR system is immediately and automatically activated whenever that stimulus is encountered again, even before it is consciously recognized as such. We therefore do not have to *think* before we *act* in dangerous situations (although we can *subsequently* reflect on what has transpired). On this basis, LeDoux distinguishes two aspects of fear-anxiety. The first is the one that we have just described, which is mediated by a "quick and dirty" (LeDoux, 1996, p. 163) pathway from the amygdala to the PAG and excludes cortical consciousness altogether. The existence of such a pathway has important implications for psychoanalytic clinicians, in that it explains how it can happen that patients feel anxious in certain situations *without knowing why* (i.e., on the basis of "repressed" past experiences or other unconscious associations). The second, slower pathway includes the cortical tissues of the hippocampus—which is of critical importance for episodic memory (see chapter 5). This enables the autobiographical self to consciously *recognize* what has happened and to deliberate reflexively upon it. This "extended consciousness" pathway also links the FEAR system to the *executive systems* of the brain, which leads us to the second of LeDoux's important points.

THE TAMING OF AFFECT

Once connections of the type just described have been made, they are *indelible*; nothing can extinguish the fact that the noxious object, place, or situation has been incorporated into the FEAR system's lexicon of "dangerous" things (Le Doux, 1996, pp. 250–252). The fact that such things cannot be forgotten has definite evolutionary biological advantages. However, they can also be *mal*adaptive. An object, place, or situation that was dangerous in early childhood—when the subject was helpless and vulnerable—might not be equally dangerous or even dangerous at all to the adult. Under these circumstances, it would be highly inexpedient if repeated or renewed exposure to the once-feared object continued forever to unleash full-blown anxiety attacks (feelings of overwhelming trepidation, freezing, fleeing, hiding, palpitations, rapid breathing, etc.).

For this reason, although the link between such objects and the FEAR system is indelible, the *output* of the system can nevertheless be **inhibited**.[10] In other words, although the association still exists unconsciously, its influence on extended consciousness and voluntary behavior is damped down or even blocked entirely. As mentioned in chapters 1 and 3, the apparatus for such inhibitory control is located in the frontal lobe—in particular, in the ventromesial and orbital frontal areas (Figure 4.8). When the *outward manifestations* of fear-anxiety reactions are extinguished in laboratory animals (through behavior-modification techniques), what functional brain imaging reveals is that the FEAR system *continues* to be highly activated, to almost the same extent as it is in animals displaying full-blown fear-anxiety responses. What differs dramatically between the two groups is that the frontal lobes are *concurrently* highly activated in the fear-inhibited group. As we discussed in chapter 1, the extent of frontal-lobe development is what distinguishes us humans most from other mammals. This is also what most distinguishes the brain of the adult human from that of the child. The frontal lobes develop rapidly during the first few years of life and continue to do so until late adolescence. These neuroanatomical facts explain the enormous differences with respect to flexibility and degree of emotional control that distinguish the human adult from the child and from other mammals. The implications for some forms of psychopathology are again obvious.

Presumably, similar mechanisms exist for *all* the basic-emotion command systems, although they have been less thoroughly studied. Unbridled affective responses of the types released by

[10] For an encyclopedic and multidisciplinary survey of research relevant to this topic, written especially for psychoanalytic readers, see Schore (1994).

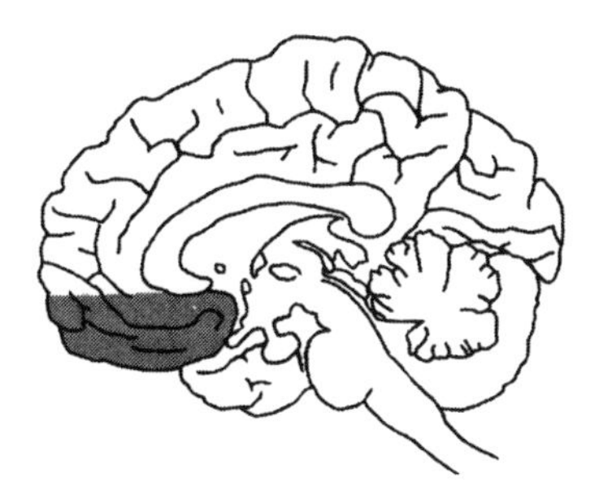

Figure 4.8
Ventromesial and orbital frontal areas

the SEEKING, RAGE, and PANIC systems are no less maladaptive, in principle, than overwhelming FEAR reactions. The balance between these primitive instinctual mechanisms and the voluntary programming, regulation, and verification of action (chapter 1) contributed by frontal-lobe mechanisms, therefore, seems to provide a direct parallel with the balance between what psychoanalysts classically referred to as the "id" and "ego" mechanisms. Damage to these frontal mechanisms is what led Harlow (1868) to observe in the case of Phineas Gage (described in chapter 1) that "the equilibrium or balance, so to speak, between his intellectual faculties and animal propensities, seems to have been destroyed." However, by the same token, the inhibitory brain mechanisms just described can also be *over*active—and this might provide the neural basis of some pathological varieties of repression and other forms of inhibition. Here, once again, there are very fertile fields for future interdisciplinary research.

In discussing such issues, we are leaving the terrain of the basic-emotion command systems themselves and moving deeper into the functional architecture of the *experience-dependent* systems of the brain.

CHAPTER 5

MEMORY AND PHANTASY

In the previous two chapters we discussed brain mechanisms that respond to the two major sources of stimuli with which the mind has to contend, and we described a number of links that have evolved between these two sources of stimuli and the brain's motor output mechanisms, most of which are probably innate. In this chapter we discuss the relationships between these two classes of knowledge that are created during the living of an *individual* life. These links enable the subject to fine-tune his or her need-satisfying activities in relation to the idiosyncrasies of the specific environment into which he or she is born. The survival value of such memory systems is obvious. Although the *content* of the memory systems is unique to each individual, memories are organized according to a regular, standard pattern. This "standard" pattern of organization of human memory, across a number of subsystems, is the main theme of this chapter. We begin with an introductory tour of these subsystems before moving on to some related topics.

The term "memory" covers many different mental functions. Sometimes we think of memory as the act of *remembering*. This aspect of memory is *reminiscence*, the bringing to mind of some previously learned fact or experienced event. At other times, the term "memory" refers not to the process of bringing stored knowledge to mind but, rather, to the stored knowledge itself. This meaning of "memory" denotes the part of the mind that contains *traces of influence* from the past that persist in the present. The term "memory" is also used in connection with the *process* of acquiring knowledge—that is, the process of learning or memorizing.

Because the function of memory covers so many different things, cognitive scientists today divide it up into a number of component functions,[1] several of which are discussed in this chapter.

ENCODING, STORAGE, RETRIEVAL, AND CONSOLIDATION

Three stages in the processing of memory are frequently referred to in the specialist literature (see Figure 5.1).[2] The acquiring of new information is called **encoding**, retaining the information is described as **storage**, and bringing the information back to mind is **retrieval**. Placing the functions of memory in the sequence of encoding, storage, and retrieval provides a simple way of dividing

---

[1] There are several books that provide accessible overviews of this literature: Larry Squire's (1987) *Memory and Brain* emphasizes the neuroscience issues—although his book is starting to date now. Daniel Schacter's (1996) *Searching for Memory* emphasizes the cognitive issues. Alan Baddeley's (1997) *Human Memory: Theory and Practice* also provides a comprehensive, but rather technical, review of the cognitive literature.

[2] Box-and-arrow diagrams, as in Figure 5.1, never reflect the reality of a mental function, nor are they meant to. We use such diagrams to *simplify* our metapsychological picture of the "mental apparatus" (see chapter 2) and mainly for *didactic* purposes. In reality, the component functions of the mind fluidly interrelate with each other in far more complicated ways than any box-and-arrow diagram can convey.

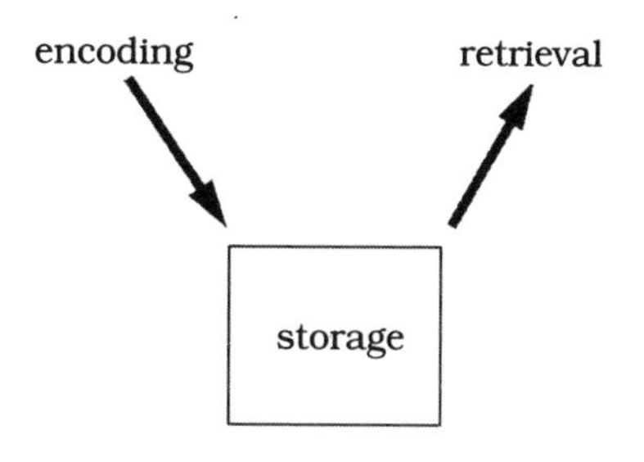

FIGURE 5.1
Encoding, storage, and retrieval

up the topic. However, these three concepts by themselves do little justice to the complexities of the neurobiology of memory.

Our simple classification already runs into trouble as soon as the concept of **consolidation** is introduced (see Figure 5.2). This has become an important concept in memory research, as it seems to cast significant light on how memory is actually organized in the brain. Compelling evidence for the existence of consolidation first came to light from studies of the way in which memory breaks down following brain damage.

It is almost invariably the case that memory is not affected globally following brain damage. It hardly ever happens that someone's memory is completely destroyed—in fact, when a

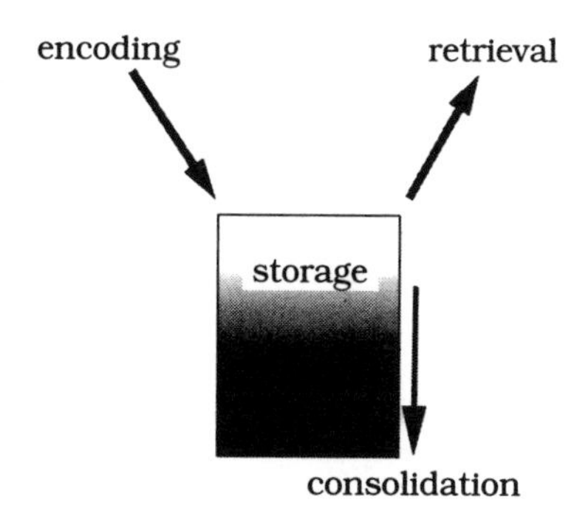

FIGURE 5.2
Encoding, storage, retrieval, and consolidation

patient suffers from a total amnesia, we readily entertain a diagnosis of hysteria. The neuropsychological reality is that particular *aspects* of memory are vulnerable to brain injury or disease, whereas other aspects are almost indestructible. The most vulnerable memories are the recent ones, the memories of events that occurred (or facts that were learned) just a few hours, days, months, or weeks before the brain was damaged. As a rule, the more remote the memory, the less likely it is to be disrupted by neurological pathology. The discovery of this *temporal gradient* (as it is often called) is attributed to Théodule Ribot, in the 1880s, and it is therefore described as Ribot's law. It is perhaps surprising that the most recent memories—the freshest ones—are the most vulnerable, while the more remote memories are the most durable. The fact that this is indeed the case suggests that something *entrenches* memories over time. This "something" is the process of consolidation. Memories are continually being consolidated to deeper and deeper levels of storage. At this moment, very little consolidation of the material you are reading is taking place; tonight (while you are sleeping) a fair amount of consolidation should take place; and over the next few days, weeks, months, and years the process of consolidation will continue. Consolidation is perhaps best conceptualized as an aspect of the encoding stage of the memory process that continues into the storage stage.

SHORT- AND LONG-TERM STORAGE

Our diagram of memory processes is complicated further by the fact that the storage aspect must be divided into **short-term** and **long-term** components (Figure 5.3). The distinction between short- and long-term memory is probably the most important division within the memory systems of the brain. It is also an important source of terminological confusion. For many people,

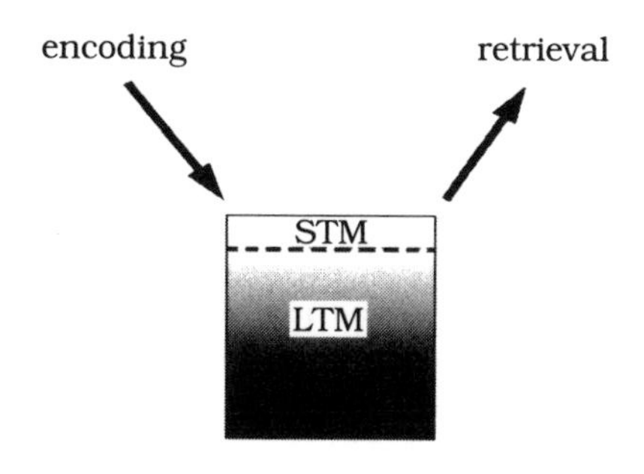

FIGURE 5.3
Short-term memory (STM) and long-term memory (LTM)

the expression "short-term memory" denotes memories laid down over the last few hours or days. People say, "My short-term memory is so bad, I can barely remember what happened yesterday!" In technical usage, however, we would say that such people are describing a difficulty with their *recent* memory. In technical parlance, "short-term memory" (STM) refers to information that is in your consciousness *right now*, derived from events that probably occurred just a few seconds ago. Both recent and remote memory are aspects of "long-term memory" (LTM). If a patient cannot remember what happened yesterday, therefore, there is something wrong with his or her *long-term* memory. Long-term memory begins a few seconds ago. It is partly as a result of this ambiguity (among other reasons) that the term "short-term memory" is falling out of use in cognitive science these days and is being replaced by the terms **immediate memory** and, increasingly, **working memory**.

Short-term (or immediate or working) memories, then, are memories of events (or facts) that you are holding in mind *at this moment*. They may be there because you have just learned them or experienced them (because they have just happened *to* you) and therefore have not yet disappeared from your consciousness. Or they may be there because you are *actively* holding them in mind, wanting to keep them in conscious awareness, or because you actively brought them to mind (from LTM). This reveals that short-term memory has an active and a passive aspect to it. We [MS, OT] tend to use the term "immediate memory" to refer to the passive (perceptual) aspect of short-term memory, and we reserve the term "working memory" for the active (cognitive operation) aspect. As we use these two terms, then, they refer, respectively, to the externally and internally generated *current contents of consciousness*.

The contents of consciousness are held in what cognitive scientists call a "**buffer**," where it can (if we wish it to) be sustained by continuous rehearsal. The mechanism of working memory can hold information in this buffer for as long as you like (until you fall asleep, that is!). For this reason, the buffer of short-term memory can be regarded as the medium of our *extended consciousness* (described in chapter 3). Cognitive scientists use terms like "ultra-short-term" and "iconic" memory to denote the momentary effects of external perceptual stimuli in *core consciousness*.

As you read these lines, the information they contain is held in the buffer of your short-term memory. Within a few seconds, as you read on and have to encode more information, the lines you read a few moments ago will have to be transferred out of the STM buffer, to make way for new information. This is due to the fact, referred to in chapter 3, that although you can keep information in mind for an extended period of time, the STM buffer of consciousness has a very limited *capacity* (roughly seven units of information).[3]

This leads to the question of what happens to the material that has been bumped out of the buffer. You cannot encode and store absolutely everything you experience. Indeed, attentional

[3] Breuer and Freud noted as early as 1895 that consciousness and memory are, in this sense, mutually exclusive (Freud, 1895d).

mechanisms exclude a good deal of information at even the perceptual stage. The mechanism of consolidation continues this sifting process within the memory-storage systems themselves. Consolidation is therefore not only a process of entrenching what is kept in memory, it is also a process of getting rid of the memories you do *not* want to keep. This introduces an important distinction between passive and active forgetting (trace decay vs. "repression": see Anderson & Green, 2001, for some recent findings). We return to this issue later in the chapter.

Earlier, we suggested that a fair amount of consolidation of what you are reading here will occur tonight. Many neuroscientists believe that the function of sleep (and of REM, or dreaming sleep, in particular) is intimately bound up with the process of consolidation. This theory is controversial. One especially mischievous claim is that dreams are the "dustbins" of memory (Crick & Mitchison, 1983). According to this view, memories are consolidated during REM sleep, and the ones that are selected for erasure appear briefly in your dreams, on their way to oblivion. Hence, the most *irrelevant* facts of the day turn up in your dreams, and this is why dreams are so readily *forgotten*. Most psychotherapists hold quite different views on the nature and function of dreaming. We return to these issues in chapter 6.

THE PHYSIOLOGY OF CONSOLIDATION: CELLS THAT FIRE TOGETHER . . .

The physiology of short-term memory is not very well understood, but neuroscientists agree that it differs radically from that of long-term memory. Short-term memory appears to involve reverberating circuits—groups of interconnected cells firing together in closed (self-reactivating) loops. The maintenance of the firing pattern is the holding in mind of the information. Once a particular reverberating circuit has been established, it is more likely to become activated *again* for the reason proposed by Donald Hebb (1949) that "cells that fire together, wire together"; this is known as **Hebb's law**. The process of "wiring together" is what transforms short-term memories into long-term ones. It seems to involve a two-stage process. Initially, the cell changes are purely *physiological*, in that the synapses connecting the cells in the circuit become more "permeable" (i.e., their thresholds decrease, making the cell more likely to fire in response to stimuli at synapses that caused them to fire previously). This in turn sets off a second, more permanent, *anatomical* process. The continual firing of cells at certain junctions activates in the cells genetic mechanisms that promote the growth of further synapses at those junctions. (For a more detailed account of these mechanisms, see Kandel, Schwartz, & Jessell, 2000.) Thus the cells *literally* grow and "wire together" at constantly activated junctions.

This relatively recent discovery, which won Eric Kandel the 2000 Nobel Prize for physiology and medicine, has very important implications for our understanding of memory. It demonstrates that temporary, reverberating circuits have a permanent, trophic effect on the cells involved, producing an increased density of neural tissue. This trophic effect is **activity-dependent**, and it continues throughout life.

FORGETTING, REPRESSION, AND INFANTILE AMNESIA

The process of "wiring together" inevitably has a flip side: If the process is activity-dependent, then what happens if a particular circuit falls into *dis*use? What happens to synapses that are *not* active? The answer is that they *atrophy*; they literally *die*. This "use it or lose it" rule plays an important part in early brain development. We are all born with billions more synapses than we need. These synapses represent the *potential* connections between neurons that *might* be needed to create internal maps and models of the world in which we find ourselves. In a sense, they represent all the possible worlds we might find ourselves in. The *actual* environment we are born into results in only a subset of these connections being activated. These particular connections are then strengthened, and the ones that are not used fall by the wayside. This process is commonly referred to as neuronal "pruning."

But the process does not end in early childhood. Although that is when the vast bulk of excess neural tissue is shed, the "use it or lose it" principle continues to operate throughout life. As a result, connections that may have been activated frequently in childhood (and therefore preserved) can subsequently fall away at later stages of development, for the simple reason that they are no longer required. This fact forms the basis of an interesting argument against the psychoanalytic theory of "infantile amnesia."

This argument is usually formulated in the following way: People do not rely on the same memory circuits in adulthood as they did in childhood, because their circumstances change so radically. Since the childhood memories are no longer used, they atrophy. Infantile amnesia is therefore a simple matter of memory decay—the disintegration of ancient connections that have fallen into disuse. There is therefore (it is argued) no need to postulate an active "repressing" force to explain the universal inability to recall the events of earliest childhood, and there is no point in trying to "recover" them.

There are several substantial problems with this claim, two of which are worth mentioning briefly. The first is the fact that *conscious* and *unconscious* remembering are two entirely different things. The *activation of a memory trace* is not at all synonymous with *conscious remembering*. The fact that you are not consciously aware and mindful of the events of early childhood does not therefore mean that the traces they left are not constantly being activated. On the contrary, it is quite likely that the networks that survived the great pruning processes of early childhood serve as *templates* around which all later memories are organized. These deeply consolidated "trunk" circuits would be activated on a very regular basis, even if the events that forged them in the first place are not consciously brought to mind in the process—and even if the events that forged them *cannot* be brought to consciousness any longer. This introduces some important points about the functional architecture of human memory, to which we turn our attention in the next section. For now, we will just mention that the distinction between *conscious* and *unconscious* memory mechanisms is very well established in contemporary neuroscience (see chapter 3). Nobody doubts that a long-term memory trace can be activated without an attendant experience appearing as a conscious reminiscence. In fact, most memory processes take this form. Such memory processes are described as **implicit**. When a long-term trace is activated *and* brought to conscious awareness (i.e., when, in addition to being activated, it becomes available to the temporary "buffer" of working memory, mentioned above) we say that it has been rendered **explicit**. (A good example of this distinction was provided in chapter 4, in the section "The Taming of Affect," in connection with FEAR memory.) The technical terms "implicit memory" and "explicit memory" in contemporary neuroscience are synonymous with the older terms, "unconscious" and "conscious" memory, respectively.

A second reason to question the claim that early childhood memories are simply "forgotten" invokes Ribot's law, which states that the oldest memories are the most robust memories. Any account of infantile amnesia must explain why it violates Ribot's law. In psychoanalysis, the explanation is that early childhood memories *are* very robust; they only *appear* to be forgotten, but in fact are just unavailable to conscious awareness. The question then becomes: Why are they not available to conscious awareness? (The answer, in psychoanalysis, is repression.) It is not clear how the alternative account explains this violation of Ribot's law.

MNEMONIC DIVERSITY

Freud is supposed to have said that once a memory is laid down, it can never be forgotten. This was not *really* his opinion, but he certainly did emphasize the remarkable persistence of memory.[4] And long-term memory is indeed a very durable thing.

The reason that long-term memories are so enduring is that they are generally encoded in several places—in a sense, memories are "everywhere" in the brain. From the earlier discussion on the nature of connections between cells, it should come as no surprise to learn that memories have a widely distributed anatomical representation. For this reason, there is a great deal of redundancy in the mnemonic process. Memories involve connections between vast assemblies of neurons, and removing one or another piece of the assembly will not get rid of the whole. It may degrade it slightly, but it is very difficult to obliterate an entire network. (By the same token, degraded traces can be "reconstructed," although the reconstructed version may not always be accurate; see below.) A second, related reason why long-term memory is so robust is that *memories are encoded in more than*

[4] What Freud actually said was this: "Perhaps we ought to content ourselves with asserting that what is past in mental life *may* be preserved and is not *necessarily* destroyed. It is always possible that even in the mind some of what is old is effaced or absorbed . . . to such an extent that it cannot be restored or revived by any means; or that preservation in general is dependent on certain favourable conditions. It is possible, but we know nothing about it. We can only hold fast to the fact that it is rather the rule than the exception for the past to be preserved in mental life" (Freud, 1930a, pp. 71–72).

one way. There is a plurality of memory subsystems, not just one "filing cabinet." So even if one "file" is lost or degraded, much of the information it contained may be stored elsewhere, in different ways, in other "files."

We now introduce some of the better-known "filing cabinets" of human memory. There is some controversy in the field about whether these represent *entirely* independent categories, but this classification system is very widely used and is likely to remain useful. The categories may be visualized as subsystems of the "storage" component of memory depicted in Figures 5.1 and 5.2.

SEMANTIC MEMORY

Semantic memory is "a network of associations and concepts that underlies our basic *knowledge of the world*—word meanings, categories, facts and propositions, and the like" (Schacter, 1996, p. 151, emphasis added). This knowledge is stored in the form of third-person information of the kind that one might find in an encyclopedia. It comprises bits of objective information about the world and its workings—facts such as "dogs have four legs" and "London is the capital of Britain." There is nothing "personal" about semantic memory, in the sense that it does not represent *experiences*. It contains information that we typically share with other members of our society, especially our peer group. However, it also stores *objective* personal information—such as "I was born on the 17th of July 1961" and "I live in Bangor, Wales." Much of our semantic knowledge is encoded during the elementary-school years of childhood, but most of it is acquired even earlier than that. One should remember that semantic memory includes much "general knowledge." Indeed, we often forget that we once had to *learn* it. For example, semantic memory contains grammatical rules of language, the knowledge that objects drop when you let go of them, that cups break but balls bounce, that leaves blow in the wind. When your hands dart out to catch a falling cup in anticipation of the fact that it might break, then that movement is based on *memory*; you reach downward because you know from endlessly repeated experiences what is likely to happen. The habitual hand movements themselves (in this example) are classified under the heading of "procedural memory," a sort of "bodily" memory we discuss separately below; but the *abstract rule* "cups can break when they drop" is encoded in semantic memory.

Categories of knowledge and perception

Semantic memory can be divided into several subcomponents, so there are specific aspects of our semantic memory that can be disrupted in relative isolation. This property of semantic memory is known as **material specificity**. The rules of language, mathematical rules, the knowledge of the shapes, and the behavioral patterns of various categories of object are stored in different networks of the brain and are therefore vulnerable to being damaged separately. Much of the difference between the mental functions of the left and right cerebral hemispheres is dependent on material specificity (see chapter 8). Material specificity is, to some extent, dependent on **modal specificity**.[5] For example, circuits in the medial occipito-temporal part of the cortex, especially on the right, categorize information that enables us to recognize individual *faces*, and circuits on the lateral convexity

[5] "Modal" specificity refers to information that is confined to a concrete *perceptual* modality (e.g., vision or hearing). "Material" specificity refers to information that is confined to a particular *abstract* category (e.g., verbal vs. spatial).

of the left temporal lobe (and adjacent parts of the parietal and occipital lobes) categorize information that enables us to retrieve specific *names*.[6] The "face" circuits encode visual-specific images, whereas the "name" circuits encode auditory-specific images. However, categorical knowledge about faces and names (and the connections between them) is also stored and classified *abstractly*. To the extent that memory networks are encoded as concrete, modality-specific images rather than abstract, material-specific connections and categories, neuroscientists tend to classify them under the heading of *perceptual* rather than memory mechanisms (see below). The abstract connections between objects (or their properties) are generally classified as semantic memories.

Anatomy of semantic memory

Since semantic memory is concerned with "objective" facts and represents the world from a "third-person" point of view (even information about your self, such as "I was born on the 17th of July 1961"), it should come as no surprise to the reader—at this point in our text—that it is encoded in the exteroceptive *cerebral cortex*. The network of associations and concepts that comprise semantic memory takes the form of a "directory" of connections between the concrete images that are represented in modality-specific cortex (see Mesulam, 1998). These directories can, therefore, to a large extent, be "localized" in the cortical "association" areas that link the various unimodal cortices with one another

[6] It is important to note that these regions do not contain the entire memory trace of, for example, an individual face or name. Rather, critical *nodes* of such circuits are to be found in the regions in question, which results in the psychological function becoming heavily degraded when these vulnerable regions are damaged.

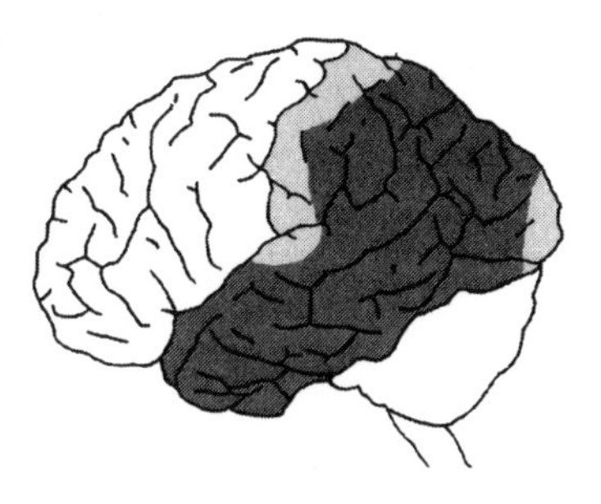

FIGURE 5.4
Posterior "association" cortex

(Figure 5.4). This applies especially to the posterior temporal and inferior parietal regions, which form the hub of the functional unit of the brain that Luria (1973) called the unit for receiving, analyzing, *and storing* information (see chapter 1). However, as previously stated, the reader should not mistake these nodal points in the associative networks for the networks themselves. The memory traces *themselves* are very widely distributed in the cerebral cortex, as they must of necessity include all the concrete, unimodal images that the semantic directories link up.

The remembered present

The ambiguity at the borders between semantic memory and perception is reflected in some curious anomalies in our clinical categorization of patients with damage to these regions of the brain. For example, although patients who are unable to remember names tell us that there is something wrong with their "memory," we (neurologists and neuropsychologists) consid
them to have something wrong with their "language" fu
Accordingly, we classify these patients under th
heading of "**aphasia**" ("anomic aphasia," to b
than "**amnesia**." (For a detailed exegesis of

see chapter 8.) Likewise, an inability to recognize familiar faces is classified as a perceptual disorder (an "**agnosia**"—"prosopagnosia" to be precise) rather than a memory disorder (an amnesia). Similarly, the inability to recall the movement one makes to catch a falling cup is a disorder of skilled movement ("ideomotor or ideational **apraxia**") rather than an amnesia. *All* the aphasias, agnosias, and apraxias are, in reality, disorders of *memory* (in its broadest sense), but we classify them as disorders of language, perception, skilled movement, and so forth. This is, in part, because these categories of knowledge are so overlearned that we overlook the fact that we ever did learn them.

As a result, much of what we take for granted as "the way the world *is*"—as we *perceive* it—is in fact what we have *learned* about the world—as we *remember* it. This is best demonstrated by the fact that the way the world "is" can suddenly change—often dramatically—for people whose brains are damaged. As a result, and not surprisingly, some patients have great difficulty recognizing that it is them*selves* rather than the *world* that has changed. This can also be demonstrated neurodevelopmentally. It is possible to "engineer" a cat that cannot *see* horizontal lines by depriving it of this type of experience in crucial developmental periods. The visual cortex of such cats becomes organized in a way that lacks horizontal information. If you confront a cat deprived from birth of horizontal visual experiences with a horizontal line (by say, putting a horizontal bar in its path), it will behave as if the object does not exist and walk straight into it. This is one class of evidence of the fact that much of what we take to be perception is in fact memory. Another example of memory-based perception is *accent*, which reflects differences between the learned features of different languages. Japanese people have great difficulty distinguishing between "r" and "l" sounds because, in the phonologically meaningful surround that their brain develops, this distinction does not (meaningfully) exist. Even if they are later put into an environment where it does have significance, they perceive the world differently—at least as far as that tiny little detail is concerned.

The title of Gerald Edelman's popular book, *The Remembered Present* (1989), captures very well what perception is about. We all automatically reconstruct the reality we perceive from models we have stored in our memories. We do not perceive the world anew every moment of the day and try afresh to discriminate recognizable objects and decipher meaningful words from the undifferentiated din of stimuli that constantly impinge on us. This, presumably, is what newborn babies have to do. We adults *project* our expectations (the products of our previous experience) onto the world all the time, and in this way we largely *construct* rather than perceive (in any simple sense) the world around us. Thus, the world of our everyday experience is doubly removed from the "reality itself" that philosophers speak of (see chapter 2)—first by the interposition of our perceptual apparatus (which is designed to *sample* and *represent* certain selected features of the world), but also by our memory (which, on the basis of past experience, organizes and transforms those selected features into recognizable *objects*).

Aleksandr Luria (the Russian neurologist we have already mentioned more than once), together with his colleague Lev Vygotsky, argued that the hierarchical arrangement of perception and memory reverses during the maturational process (see Luria, 1973, pp. 74–75). For a small infant, everything depends on the senses, and cognition is driven by concrete perceptual reality. During the course of development, however, deeply encoded and abstract knowledge derived from these early learning experiences comes to govern the perceptual processes. We therefore see what we expect to see, and we are either surprised or fail to notice when our expectations are contradicted. Experimental studies show that we frequently see things that are not there, simply because we expect them to be there. The best-known example of this is provided by the "blind spot," which is located in each eye at the point where the optic nerve enters the retina. For this reason, objectively, we have a hole in our vision (not far from the middle of the visual field) when we close one eye. Subjectively, however, this region is "filled in" with the texture, color, movement, and the like that are appropriate to what we *expect* to experience in that part of the visual field under prevailing circumstances. This is an example of what cognitive scientists call "**top-down**" influences on visual perception. (Only babies can be argued to rely almost exclusively on the **bottom-up** perceptual mechanisms.)

These facts are obviously important for psychotherapists whose daily work is concerned, perhaps above all else, with helping their patients to become aware of the internalized models that govern their life experiences—and render the present as if it were the past. It is not certain whether the findings of neuroscience in relation to the top-down influences of memory mechanisms on perception apply also to the complex relational phenomena of *transference* and the like that psychotherapists are interested in. However, it seems a reasonable working assumption that these mechanisms explain at least *part* of these more complex phenomena. (Here, once again, we have fertile soil for future interdisciplinary research; see chapter 10.)

PROCEDURAL MEMORY

Procedural memory is a kind of "bodily" memory. It is memory for habitual *motor* skills, or, more generally, *perceptuomotor* or *ideomotor* skills. It "allows us to learn skills and know how to *do* things" (Schacter, 1996, p. 135, emphasis added): knowledge about how to walk, how to stack blocks in towers, how to write, how to play the piano. As we said above, many of these skills are so overlearned that we do not normally think of them as aspects of *memory*. However, *as learned skills, retrieved when appropri-*

ate, that is what they are. They depend on the right sort of experience, and a great deal of practice. Constant repetition in the learning phase is especially important for procedural memory—which has far deeper evolutionary roots than semantic memory does. All levels of ideomotor ability, from walking through playing the piano, are skills that are learnt gradually. Skills like riding a bicycle are also *extremely resistant to decay with time*. An aphorism that is therefore frequently applied to procedural skills is that they are "hard to learn, hard to forget."

There is a degree of overlap between procedural and semantic memory, as many motor skills are encoded and stored in both procedural and semantic forms. A useful way to distinguish them is to think of the difference between one's concrete skill in *playing* a particular game and (say) one's abstract knowledge of the *rules* of that game.

The distinction between procedural and semantic memory is underscored by the fact that, following brain damage, they can break down independently of each other. It is quite common for neurological patients to lose habitual abilities but retain abstract knowledge about the skills they have lost. Accordingly, functional-imaging studies (e.g., PET and fMRI)[7] reveal that different parts of the brain are activated in procedural and semantic memory tasks. However, the parts of the brain that are activated in procedural-memory tasks do not constitute the *entire* motor system. For example, cortical motor (and ideomotor) structures in the parietal and frontal lobes are engaged in procedural *learning*. However, once a skill becomes *habitual* (i.e., more

[7] PET (positron emission tomography) and fMRI (functional magnetic resonance imaging) are techniques that reveal the relative activation of different parts of the brain by scanning the level of metabolic activity (which reflects the rate at which cells are firing) in different regions of brain tissue. Using these techniques while someone performs a particular task, and comparing the results with those for a different task, reveals differences in the parts of the brain involved in the different tasks.

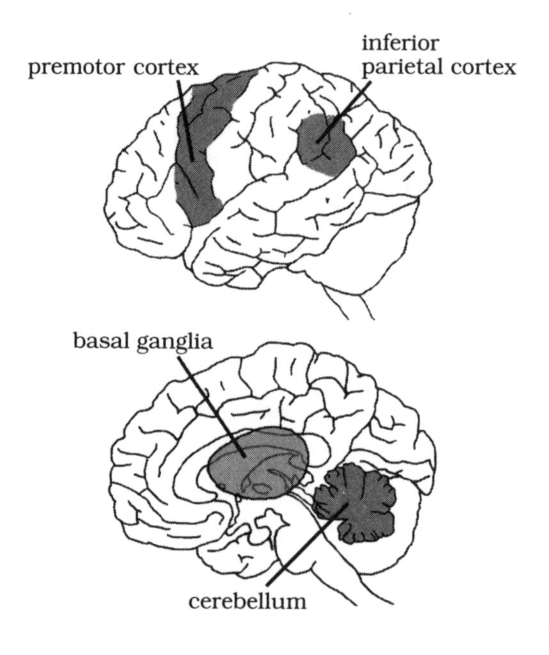

FIGURE 5.5
Regions involved in procedural learning

deeply consolidated in procedural memory), the motor program representing it is progressively consolidated into subcortical structures, involving mainly the basal ganglia and cerebellum (see Figure 5.5).

Procedural memory and the unconscious

An important feature of procedural memory is that it functions *implicitly*. Habitual behavior is executed automatically (and therefore unconsciously) almost by definition. As soon as a procedural memory is rendered explicit, it becomes something else; it is translated into a semantic or episodic form (episodic memory is discussed in the next section). For example, it is widely known that game-playing skills can improve without a commensurate increase in abstract and explicit knowledge of how one is *supposed* to move when playing the game. This knowledge, which *instructors* of (say) tennis or golf have, is *not* acquired merely by practice at the procedural aspects of playing the sport. Many extremely competent players have no knowledge of the detailed movements required to execute a particular shot. In fact, a well-known form of gamesmanship in such sports is to *ask* your opponent how exactly they hold the racquet, or what position their elbow takes during a stroke. The experienced player knows that explicitly thinking about what was previously a well-rehearsed movement often causes a steep decline in performance levels. In contrast, top sportsmen describe their best, or peak, experiences as happening when they are "in the zone"—a situation of complete automaticity in which they do not think *at all* about how to, say, execute a stroke, and the racquet or club seems simply an extension of their body (see Gallwey, 1986).

Typically, procedural memories will be associated with both semantic *and* episodic memories. That is, the same experiences will have been encoded in different ways simultaneously—as a set of experiential episodes, as a set of abstract facts, and as a set of habitual responses. This is a manifestation of the redundancy in memory that we mentioned before. As a result, it is quite possible (indeed, commonplace) for a person's behavior to be heavily determined by influences and events of which that person is totally unconscious.

This is obviously relevant to some of the phenomena that psychotherapists deal with. It adds another dimension to what we said about "transference" and the like in relation to perceptual memory. Transference clearly encompasses aspects of procedural memory too. It is uncertain to what extent this applies to other phenomena of interest to psychotherapists—such as the "bodily memories" that some traumatized patients display.

However, as we discussed previously, some automatic emotional behaviors (like unconscious FEAR reactions to conditioned noxious stimuli) certainly do seem to function very much like procedural memories. Perhaps future interdisciplinary collaboration between psychotherapists and neuroscientists will help us to differentiate more precisely among "procedural" memory subsystems.

EPISODIC MEMORY AND CONSCIOUSNESS

Episodic memory involves the literal "*re*experiencing" of past events—the bringing back to awareness of previous experiential episodes. This is what most of us think of as memory proper. When we say "*I remember* . . . [anything]," we are speaking of an *episodic* memory. According to Schacter (1996), the episodic-memory system "allows us *explicitly* to recall the personal incidents that *uniquely* define our lives" (p. 17, emphasis added). The emphasis here falls on the twin facts that these memories are intrinsically *subjective* and that they are intrinsically *conscious* (hence the *I* and the *remember*).

Why should our memories of personal life events necessarily be conscious? Herein lies an important problem. These memories are conscious because they involve the reliving of past *moments of experience*. We know from chapter 3 what these moments of experience consist of: they are momentary couplings of states of the self with concurrent events in the outside world—and we know that *consciousness* (or "core consciousness") is both the medium and the message of such couplings. Episodic memory, then, constitutes the essential tissue of the "autobiographical self" (see chapter 3). Extended consciousness is "extended" precisely because it *extends* the quality of consciousness backward onto *past* self–object couplings. It involves the reliving of past moments (or past self–object "units") of core consciousness.

But does that really mean that autobiographical knowledge is necessarily conscious? Psychotherapists routinely report that their patients "recover" memories of personal life events of which they were previously unconscious. Were these memories not then previously encoded as "episodes"? Did they previously exist only as semantic beliefs and procedural habits? If that is so, all so-called *recovered memories* would in fact be *reconstructed* memories, in the sense that they would be made from raw material that was not, in itself, "episodic." On the one hand, it certainly seems plausible that a personal episode can leave a neural trace (a self–world connection) that links two veridical representations (a state of the self with concurrent events in the world) and only becomes conscious once the *link* (as opposed to the *representations* themselves) is activated again. And yet, it is questionable whether a *state* of the self can be "represented" without necessarily being "reactivated." In other words, states of the SELF might be intrinsically conscious. (One cannot say the "I" in "I remember . . ." without simultaneously *being* it.) The *sense* of self (of "I was there . . ."; "it happened to me . . .") appears to be *necessarily* conscious. This implies that although *external events* can be encoded unconsciously in the brain (as semantic, perceptual, or procedural traces), the episodic *living* of those events apparently cannot. Experiences are not mere traces of past stimuli. Experiences have to be *lived*. It is the *reliving* of an event as an *experience* ("I remember . . .") that necessarily renders it conscious. And it is the sense of self (of "being there") that combines the traces into an experience. This is another way of saying what we said in chapter 3 in relation to consciousness in general: it is the SELF that *binds* our fragmented representations of the world into unified, lived *experiences*. The *link* in a self–world coupling is therefore the SELF itself.

Thus we seem to have rediscovered, from a neuroscientific standpoint, the obvious fact that what *we feel* about our experiences is what renders them susceptible to "repression." Even though we may have a perfectly good semantic, perceptual, or procedural record of an event, the multiple exteroceptive traces of that event have to be brought back into concurrent connection with (and by) the sentient, feeling SELF if the event is going to be consciously relived (i.e., remembered episodically). Anything that impedes such connections can banish a memory from extended consciousness.

All of this suggests that when psychotherapists speak of unconscious memories of personal events, what they are really referring to is something that the stored memories of the events in question *would be like* if they *could* be reexperienced. Unconscious memories of events (unconscious episodic memories) are "as-if" episodic memories. They do not exist *as experiences* until they are reactivated by the *current* SELF. In the interim, they only exist, as such, in the form of procedural and semantic traces (habits and beliefs).

Anatomy of episodic memory

The structures that are most important for episodic memory are quite different from those that serve semantic and procedural memory. Episodic memory involves conscious activation (i.e., arousal by the core brainstem structures discussed in chapter 3) of stored patterns of cortical connectivity (i.e., facilitated synaptic networks) representing previous perceptual events.[8] The directories of such links between the stored cortical patterns and various states of the brainstem SELF seem to be encoded, above all, through the **hippocampus**. The hippocampus is a folded

[8] Conscious *thoughts* are "perceptual events" too and can also be reactivated.

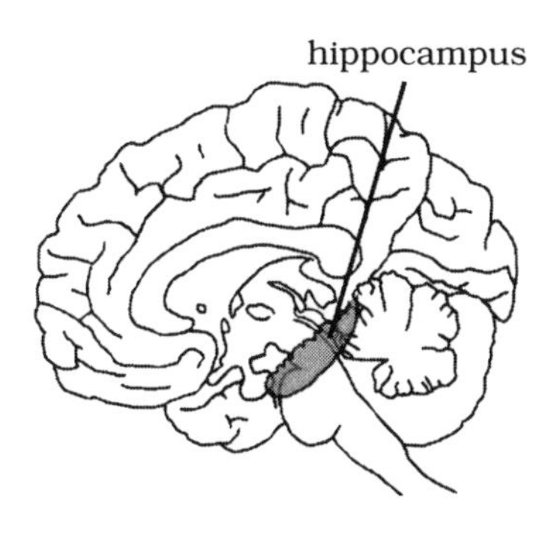

FIGURE 5.6
The hippocampus

piece of primitive cortex that lies on the inner surface of the forebrain within the temporal lobe (Figure 5.6). It is densely interconnected with a group of other structures loosely termed the "limbic system" (see chapter 1).

It is a matter of no small importance for understanding episodic memory that the network of structures that comprise the limbic system was first identified (by James Papez in the 1930s: see MacLean, 1949) not in connection with the functions of *memory* but, rather, in connection with *emotion*. This underscores the fact that episodic memories are not simply stored but, rather, *lived*. The essence of episodic memory is that it is *conscious*, and, as we learned in chapters 3 and 4, the essence of self-generated states of consciousness is that they are intrinsically *emotional*. This is why we say that consciousness is both the medium and the message of episodic memory: we retrieve events in an episodic form in order to remember what they *felt* like.

The effects of hippocampal lesions Patients with bilateral damage to the hippocampus are not unconscious. Core consciousness is completely intact in these cases. What they lose is a crucial component of *extended* consciousness: the ability to extend conscious awareness to the neural traces of past events (see chapter 3). *The traces themselves are still there* (implicitly, in procedural and semantic form), but these patients are unable to consciously (explicitly) *revive* them. For this reason, the behavior of these patients is still *influenced* by past experience; all that they lack is the ability to reflect consciously upon the experiences. A famous case of Claparède's is often cited in this connection (Claparède, 1911). Claparède concealed a pin in his hand when he greeted the patient, pricking her hand as he shook it. When he next attempted to greet the patient, she withdrew her hand, *even though she had no conscious recollection of ever having met Claparède before.* The event of the meeting had disappeared from her memory, but its effects remained. This is an example of the dissociation between episodic and procedural memory. When asked why she refused to take Claparède's hand, the patient explained that "one has the right to withdraw one's hand" (or something to that effect), thereby demonstrating the dissociation between episodic and semantic memory. She knew what to *do* (procedural memory), and she recalled relevant *abstract facts* (semantic memory), but she was unable to bring the appropriate *actual experience* (episodic memory) back to mind.

A further distinction needs to be drawn here. Claparède's patient would have been able to recall the actual experience of her hand being pricked *if it had occurred a long time before the onset of her brain damage.* Patients with hippocampal lesions lose (primarily) the ability to recall events that occur *after* the damage. This demonstrates that the hippocampus is centrally implicated not so much in the *retrieval* of past experiences as in the *encoding* of experience in an explicitly retrievable form (see Figure 5.1). It appears as if the role of the hippocampus is to *create* the (self–object) directory links referred to above. Inability to consciously recall personal events that occur after brain damage is called **anterograde amnesia**. Difficulty remembering such events that occurred prior to the damage is called **retrograde amnesia**. Typically, the episodic-memory cut-off point does not coincide *precisely* with the moment the brain was damaged; the period of amnesia for personal events usually extends some way beyond the onset of the damage. This reflects Ribot's law and is attributable to the fact that recent memories (those encoded just before the onset of amnesia) are not yet deeply enough entrenched to withstand the effects of hippocampal damage. The retrospective erasure of prior memories provides striking evidence of the existence (and importance) of the process of *consolidation* discussed above, and of the involvement of the hippocampus in this ongoing dynamic process.

The role of the hippocampus in episodic memory is also evident from the effects of hippocampal *stimulation*. Just as hippocampal lesions *deprive* perceptual traces of a sense of "I was there" or "That happened to me," so hippocampal stimulation can *produce* an artificial sense of "I was there" or "That happened to me." This is the presumed physiological basis of the *déjà vu* phenomenon, of some forms of hallucination (e.g., in complex-partial seizure disorders—see chapter 6), and (quite possibly) of certain forms of "false memory."

"HM" No account of the neuropsychology of episodic memory is complete without at least a passing reference to the celebrated case of "HM." Alongside Phineas Gage, HM is perhaps the most famous clinical case in the history of behavioral neuroscience. He suffered from an intractable seizure disorder that had its epicenter in the hippocampus (as seizure disorders often do, due to the low firing threshold of limbic neurons). In the 1950s, a Canadian neurosurgeon by the name of Scoville sensibly decided to remove the diseased hippocampal tissue that was producing the seizures. Today, this operation is still very successful in treating cases of intractable epilepsy. As a direct result of what

Scoville discovered, however, the operation that is performed today differs in one very important respect from the operation that he performed on HM. Scoville removed *both* HM's left and right hippocampi. As a direct consequence of this operation—the effects of which were later documented by Brenda Milner, a neuropsychologist colleague of Scoville's—HM never laid down another episodic memory (Scoville & Milner, 1957). This case first drew the attention of neuroscientists to the critical role of the hippocampus in memory.

HM still has good access to his premorbid memories. This means that he is only able to recall the life he led until shortly before his operation, including his childhood and early adulthood. He is, therefore, essentially still living in the 1940s. He also has normal immediate memory. Thus, he can hold onto roughly seven units of information at a time; but as soon as the information is shunted from his STM buffer into long-term memory and is replaced by a new chunk of conscious information, he can never bring the original information back to consciousness again. HM has been very extensively studied by neuropsychologists. This has provided abundant opportunity for him to demonstrate the integrity of his semantic- and procedural-memory abilities. For example, he has shown strong improvement in his scores on a wide variety of standard psychological tests, even though none of these tests *feels familiar* to him, and even though he does not recognize a single one of the professionals who have worked with him at such close quarters over all these years.[9]

Today, when the hippocampus is removed for the treatment of complex-partial epilepsy, neurosurgeons resect only *one* hippocampus, and we expend a great deal of effort to ensure that it is

[9] See Ogden (1996) for an excellent case description of HM's world. Oliver Sacks's "Lost Mariner" (in Sacks, 1985) offers a wonderfully clear description of another such amnesic patient, though there are several key differences between Sacks's patient and HM—mainly because the underlying cause of the amnesia (and hence also the precise lesion site) differ, as discussed below.

the diseased and not the healthy one that is removed. If both hippocampi are diseased, then the operation is absolutely contraindicated (on the assumption that it is better to have epilepsy than never to lay down another episodic memory again). There are also several other disease processes that preferentially affect this region. For example, this type of amnesia is frequently found after *herpes simplex encephalitis*, a viral illness that tends to attack hippocampal tissue selectively. This type of amnesia is also a common consequence of *hypoxia* (which occurs with smoke inhalation, anesthetic accidents, and near-drowning incidents, among others). Perhaps the best-known disorder causing this type of amnesia is *Alzheimer's disease*, where the pathological process very commonly begins in the region of the hippocampus and affects the hippocampus more severely than it does most other structures of the brain.

FORGETTING, REPRESSION, AND INFANTILE AMNESIA, REVISITED

A central point to grasp is that the multiplicity of LTM storage systems makes it a commonplace for experiences to influence our behavior and beliefs without us *consciously remembering* the experiences in question. The fact that you cannot explicitly bring something to mind does not mean that you do not know (unconsciously, implicitly) what happened, nor that you will not act on the basis of this knowledge. What you remember, consciously or unconsciously, depends solely on which memory systems are engaged when the memories are being encoded and retrieved. Only when the episodic-memory system is involved in the encoding (and early consolidation) of an experience can we explicitly remember that experience. If this system is not engaged, then the event will disappear from consciousness, even though its implicit effects on behavior and beliefs may well endure.

This suggests a possible physiological mechanism for (at least some forms of) repression. Facts about the forgetting of stressful experiences that are of obvious relevance for psychotherapists have come to light in recent years. The first such fact is that stressful experiences can impair hippocampal (and therefore episodic-memory) functioning. In stressful situations (e.g., those associated with activation of the FEAR system; see chapter 4), the body unleashes a cascade of events that culminates in the adrenal glands releasing steroid hormones (**glucocortico-steroids**). These hormones help us to mobilize energy where we need it (e.g., increase cardiovascular activity) and to dampen down processes that need to be inhibited in stressful situations. But, useful as they are, excessive exposure to glucocortico-steroids can also damage neurons—and hippocampal neurons in particular, since these neurons contain an unusually high concentration of glucocorticoid receptors. Schacter (1996) reviews convincing evidence to the effect that prolonged stress (e.g., in war veterans and in victims of childhood sexual abuse) results in elevated glucocorticosteroids. This is associated with various abnormalities of memory that may reflect hippocampal dysfunction. Moreover, brain-imaging studies reveal that hippocampal volume is significantly decreased in such populations. Furthermore, experiments on normal subjects show that pharmacological manipulation of steroid hormone levels can produce temporary episodic-memory impairment, even in healthy volunteers. These facts suggest that hippocampal uncoupling might well be an important component in the repression (i.e., unavailability to consciousness) of traumatic memories. These memories are not encoded in a form that leaves them accessible to subsequent conscious recall, due to hippocampal dysfunction during the traumatic moment itself.

A similar line of reasoning applies to infantile amnesia. The hippocampus is not fully functional in the first two years of life. This suggests that it is not *possible* for someone to encode episodic memories during this time period. Naturally, this does not imply that these early years are unimportant, or that we have *no memory* of the first two years of life. It implies only that the memories that we *do* encode during the very early years will take the form of habits and beliefs (procedural and semantic knowledge) rather than explicit, episodic memories. Infantile knowledge is stored as "bodily memory" and implicit knowledge about how the world works. We therefore have every reason to expect that early experience has a decisive impact on personality development (considering the evidence of "neuronal pruning" and the like; see earlier), but it seems highly unlikely that anyone can explicitly remember any event that happened to them in the first 18–24 months of life. When one is faced with an episodic memory dating from those very early years in a psychotherapeutic setting, it seems prudent to regard it as a "reconstruction" derived from sources other than episodic memory (or possibly constructions derived from *later* episodes, projected backward onto the first two years).[10] Many of the characteristics that Freud attributed to "screen memories" seem to apply here.

This has important implications for the "recovery" of repressed and infantile memories. On present knowledge, it seems reasonable to assume that *episodic* early infantile memories can never be recovered in any veridical sense. Our earliest experiences can only be *reconstructed*, through inferences derived from implicit (unconscious) semantic and procedural evidence. The same applies, to a lesser extent, to traumatic memories. It appears reasonable to assume that in some extreme cases, traumatic events simply are not encoded in episodic memory, and therefore (as in cases with structural hippocampal damage) they

[10] Our earliest childhood memories are often patched together retrospectively from photographs and parents' accounts of the events in question. The reconstructed quality of these memories is often given away by the fact that we actually *see* ourselves (from a third-person point of view) in the "remembered" episodes.

can never be retrieved as such. However, it seems likely that such events will more commonly be encoded in a *degraded* episodic form, with the result that greater effort will be required to revive them, and the final product will be more or less unreliable (patched together from vague episodic traces, and partly constructed from other sources).

RETRIEVAL DISORDERS

So far, we have focused almost exclusively on the encoding and storage stages of memory (see Figure 5.1). Although the amnesia associated with hippocampal damage takes the form of an inability to retrieve postmorbid episodic memories, this is not due to any abnormality in the brain's retrieval mechanisms *per se*. These memories cannot be retrieved because they were not *encoded* in an appropriate episodic form in the first place. The memory disorders associated with abnormalities of *retrieval* take an entirely different form.

Figure 5.7 reminds us that the hippocampus is part of a complex circuit of (limbic) structures. Nestled within the tem-

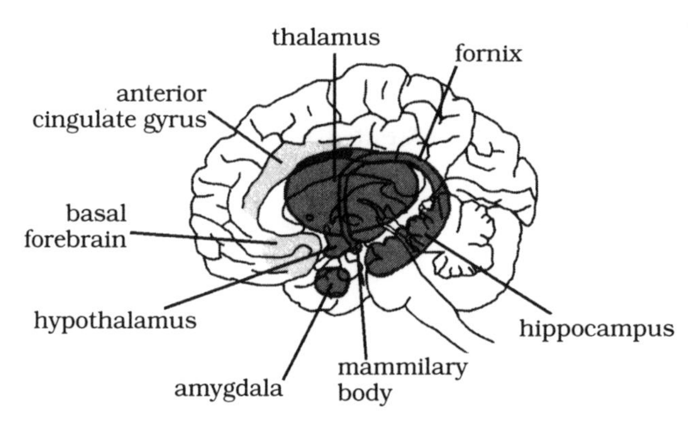

FIGURE 5.7
The hippocampus and its connections

poral lobe, which is part of the functional unit for reception, analysis, and storage of information (see chapter 1), the hippocampus may be described as the *perceptual* end of the limbic system. Through a thick bundle of axons (the fornix), which course around the diencephalon, the hippocampus projects to a group of structures nestled within the "motor" part of the brain—the functional unit for programming, regulation, and verification of action (chapter 1).

These densely interconnected structures include the *dorsomedial nucleus of the thalamus*, the *mamillary bodies*, the *basal forebrain nuclei*, and the *ventromesial frontal cortex* itself—which surrounds all these nuclei. These structures are damaged by a variety of disease processes, but perhaps most commonly by the vitamin-B deficiency associated with *chronic alcoholism* (Wernicke's encephalopathy) and by ruptured *aneurysms of the anterior communicating artery*. These pathologies produce a striking neuropsychological condition known as **Korsakoff's psychosis**. The fact that it is described as a "psychosis" immediately indicates a cardinal difference from the amnesia associated with hippocampal lesions. If you ask a patient like HM where he first met you, he will almost certainly respond that he "doesn't know" or "can't remember." Put the same question to a patient with Korsakoff's syndrome, however, and—even if you have never met the patient before in your life—he or she is likely to respond with a statement like "What do you mean, where did I first meet you . . . we've known each other for years; just yesterday, you were sitting right here and we had a drink together!" These patients do not *forget*—they *misremember*. This type of misremembering is called **confabulation**.

Confabulation is the primary distinguishing characteristic between the encoding and retrieval forms of the amnestic syndrome. Korsakoff patients do not forget, or have gaps in their reminiscences. Rather, their reminiscences contain material that does not belong there. Careful studies of these patients have revealed that their false memories don't come from thin air. Rather, they are fragments of real memories jumbled up in an inappropriate manner. (Later, we discuss some important implications of these studies for depth psychology.) The term **achronogenesis**—which means disorder of time sequence—is sometimes used to describe these memory errors. Patients suffering from achronogenesis may tell you about something that happened 10 years ago as if it had happened yesterday. It turns out that a problem of time sequencing is not the only characteristic of confabulatory errors. Another very interesting feature of these patients is their inability to distinguish between memories and nonmemories. Dreams, memories for real experiences, and daytime thoughts are frequently confused with one another (see Solms, 1997a).

An example of this phenomenon will capture the nature of the difficulty more clearly. In a standard clinical assessment of memory, one of these patients was read the following story:

> "On December 6, last week, a river overflowed in a small town 10 miles from Oxford. Water covered the streets, and entered the houses. Thirteen people were drowned, and 600 people caught colds, because of the dampness and cold weather. In trying to save a boy, who was caught under a bridge, a man cut his hand."

The patient was then asked to retell the story, and he replied:

> "There was a flood, I think it happened in Streatham [which is where he lived] . . . was it in the High Street? What happened to Jack and that shop of his in the High Street? I don't remember . . . but I remember the day I was there with him. And there was a doctor there, asking me stupid questions about my memory—he didn't realize that people who have suffered strokes can't remember!"

There are clearly some elements of the original story there, but the story rapidly becomes muddled. This includes tangential associations and confusion of memories from the past with thoughts from the present—as is demonstrated by the comment about doctors and their silly questions.

The *contents* of such patients' confabulations, and the *types* of errors they habitually make, have important implications for depth psychology. We discussed a number of striking examples in chapter 3—under the headings "Exemption from Mutual Contradiction," "Timelessness," "Replacement of External by Psychical Reality," and "Primary Process (Mobility of Cathexis)." There we pointed out that these typical features of confabulation are also the special functional characteristics that Freud attributed to the system Unconscious (Freud, 1915e, p. 187). Considering them again in the context of memory retrieval mechanisms allows us to draw some interesting conclusions about the organization of memory.

A SECOND TYPE OF MEMORY ORGANIZATION?

We have said that the hippocampus is critically implicated in the *encoding* of episodic memories, and that when it is damaged, episodic memories literally disappear (as such). However, when the diencephalic and ventromesial frontal structures that support episodic-memory *retrieval* are damaged, the memories do not disappear—they merely lose their veridical and rational organization. This is because the structures that normally systematize the retrieval process in accordance with the requirements of reality and reason (Freud's "reality principle" and "secondary process") are damaged. This suggests something very important about long-term memory and about unconscious memory systems in general. The way that long-term memories are organized and interconnected *unconsciously* may differ dramatically from the way in which we normally retrieve them consciously. The associative links that form between them may therefore be very different from what we would imagine from the point of view of the healthy, reflective ego. The features of veridicality and rationality that we normally value so highly appear to be *add-on* features that only make their appearance during the *retrieval* process, under the goal-directed control of the functional unit for programming, regulation, and verification of action (see chapter 1). In the next chapter, we shall see that this functional unit also loses its influence over our memory processes at night, while we are sleeping.

Psychoanalysts have long believed that the organizing principles of unconscious cognition are entirely different from those of conscious (and preconscious) mental life. Using the memory errors of neurological patients as a source of evidence in this regard offers us an opportunity to investigate these organizing principles from a completely different perspective to that which was used to discover the principles—that is, from the free associations of neurologically normal patients on the psychoanalytic couch (see Kaplan-Solms & Solms, 2000). Although the evidence provided here (and in chapter 3) is by no means unproblematic, and is open to criticism and reinterpretation, it suggests a novel line of approach to the investigation of unconscious memory systems. We [MS, OT] are currently conducting a systematic series of investigations of confabulatory amnestic patients from this point of view, using a range of neuropsychological techniques.

The point we are making here is simply that there is a great difference between the retrieval of information (conscious remembering) and the way in which the information is actually stored and organized unconsciously. The implicit effects that unconscious memory associations exert on our everyday cognition and behavior might be equally unexpected from the viewpoint of explicit ego functioning. It is important to remind ourselves here of a point we made near the beginning of this chapter—namely, that memory traces may be unconsciously activated all the time; one does not have to retrieve a memory explicitly in order for it to be active, and for it to influence cognition and behavior.

FORGETTING, REPRESSION, AND INFANTILE AMNESIA—AGAIN

We have just learned that frontal cortex is crucial for the retrieval of memory in a *realistic, rational*, and *orderly* way. In this context, it is notable that frontal cortex, no less than the hippocampus, is poorly developed in the first two years of life. In fact, there is a substantial growth spurt in frontal cortex at around 2 years of age, and then a second spurt at about 5 years, and frontal cortical volume continues to expand throughout the whole of adolescence. In the first few years, the level of organization of the frontal system (the unit for programming, regulation, and verification of activity) may be considered so poor that the organized retrieval process we have just discussed is, to all intents and purposes, not available to the young child. The goal-directed, rational, realistic, selective, and chronologically sequenced way of remembering that we rely on as adults is thus not characteristic of those early years of life. As a result, the reminiscences of young children are not all that different from those of adults suffering from Korsakoff's syndrome. Since the goal-directed frontal system plays an important part in *controlling* encoding and consolidation processes too, it seems highly likely that the memory traces of young children are actually *stored* differently to the way they are stored in adult brains. If something is encoded in one form, it is more difficult to retrieve it accurately in another form, thus reinforcing what we said earlier about the reconstructed nature of childhood memories. These facts, no less than those pertaining to the maturation of the hippocampus, shed important new light on the phenomenon of infantile amnesia.

All of these considerations raise the possibility that what Freud called "primary repression" or "biological repression" (i.e., the natural development of a repression barrier around the fifth year of life) may have a lot to do with normal frontal-lobe maturation. These considerations also suggest that it would be a mistake to think of repression solely in terms of *encoding* mechanisms (i.e., failure of *hippocampal* memory mechanisms). *Retrieval* mechanisms, and the *frontal lobes*, obviously also play an important part in the developmental and clinical phenomena that Freud conceptualized under the rubric of "repression." Also, in certain individuals and in certain situations, the retrieval processes run by the frontal lobes might well be selectively *biased* in the type of material that is promoted for conscious representation in the episodic-memory system. But here, too, the reader should not forget what we said earlier about failures of episodic memory: the fact that something is not *consciously* remembered does not mean that it is not remembered at all. Presumably "repressed" memories, no less than other forms of implicit memory, continue to exert a definite effect on cognition and behavior, throughout life, via the procedural- and semantic-memory systems.

THE FRONTAL LOBES, EMOTION, AND MEMORY

Chapter 4 reviewed the neurobiology of the various core emotion systems. These are involved in learning about the *consequences* of our actions—so that we are better able to predict these consequences in the future. We are also beginning to understand something of the way in which the output of these systems interacts with cognitive processes, and the manner in which it enters conscious experience. The anatomical basis for this interaction appears to be the ventromesial frontal lobes—discussed earlier, especially in relation to the case of Phineas Gage (chapter 1). It is in this brain region that the fiber pathways from the various subcortical emotion systems begin to interact with the cortical (and more cognitive) systems of the frontal lobes. This offers a mechanism by which core emotional information can access the highest-order and most sophisticated parts of the mental apparatus.

Understanding this topic may help us to resolve a long-standing problem in neuropsychology. Neurological patients such as Phineas Gage have so far proved a puzzle to the neuropsychological community. Such patients show relatively normal intelligence and, in many cases, near-normal performance on a range of tasks specifically designed to test "frontal-lobe" function. However, in spite of such apparently normal performances, they choose unsuitable friends, enter inadvisable relationships, and engage in inappropriate activities (Bechara, Damasio, & Damasio, 2000). This behaviour typically leads to financial losses, career termination, and loss of the affection of family and friends. The role of emotion—and especially that for emotion *learning*—has recently changed our understanding of the behaviour of such patients. It appears that the poor judgement- and decision-making abilities of these patients follow from an inability to use the emotion-learning systems, which provide information about the likely outcome of future decisions (see Damasio, 1994, 1996).

The literature cited above has suggested a biological basis for the substantial role of emotion in cognition, and this aspect of mental life can now be reliably assessed using the Iowa Gambling Task (Bechara, Damasio, Damasio, & Anderson, 1994). In this task, subjects are presented with four decks of cards and are asked to choose any deck, in any sequence, and then take a card from it. They win or lose money with each turn of a card.

Two decks have frequent high gains but also occasional substantial losses; sustained playing with these decks leads to overall financial loss. The other decks have more modest payouts but lead to only small and infrequent loses, so that sustained playing with these decks leads to overall moderate gain. The subjects are not told about the specific differences between the decks, but they are informed that some are "better" than others—that is, that they are more likely to win from some than from others. The game is complex, and participants do not appear, *subjectively*, to understand the contingencies of the game. Nevertheless, they can quite rapidly develop a "feeling" about which decks are good or bad; this probably derives from small activations of emotion in the seconds *preceding* the choice of a high-risk "bad" deck—when the participant is *imagining* which deck he or she might choose (see Damasio, 1994, 1996). Activation of the autonomic nervous system is the physiological correlate of this emotional experience, and it can be directly measured using changes in skin conductance (see Damasio, 1994, 1996). In other words, participants receive "advance warning" of the consequences of their actions, coded in terms of *emotion*, which allows them to avoid negative consequences (Bechara et al., 1994).

In practice, all participants start by choosing the risky decks, but neurologically normal participants (even those who regard themselves as "gamblers") rapidly shift to decks where they will accrue the smaller amounts of money over longer periods. Neurological patients with lesions to the ventromesial frontal lobes show a strong skin-conductance response *after* a bad choice has been made—which indicates that they still *feel* emotion—but they have no ability to develop the "advance-warning" effect that could alert them of a potentially poor-outcome choice. As a result they do not develop an avoidance of such choices, and they consistently lose money (Bechara et al., 1994). This inability to predict the likely emotional outcome of their actions is probably the cause of their many difficulties in everyday life.

Intuition and subjective experience

Thus, successful participants on the Iowa Gambling Task seem to acquire normal performance by using an *implicit* learning system—relying on an (emotionally mediated) "feeling" or "hunch" about which decks are good or bad, in the absence of explicit (cognitive, conceptual) awareness of the way that the decks pay out (Bechara et al., 2000). This is a reasonable definition of reliance on "intuition" or belief—that is, confidence in something not immediately susceptible to proof. Participants cannot prove, or explicitly demonstrate, *why* they are choosing the decks that they do—but they are prepared to "trust their feelings" in making the choices. In other words, participants are required to base their choices on a system that *appears* to be outside rational control, because they are asked to "feel" their way through performance on the task. In practice, their ability to perform well is far from mysterious—they are merely attending to the input from a second (affective) source of knowledge about the properties of objects. Decisions are therefore made on the basis of two sources of information—cognitive *and* affective.

As the evidence from patients with ventromesial frontal lesions demonstrably shows, the affective source of knowledge is *central* to learning and problem solving; nevertheless, it is clearly an underresearched aspect of decision making (Fridja, Manstead, & Bem, 2000). It is also, of course, a phenomenon of great interest to psychoanalysts: the analytic situation regularly requires the analyst to make judgments on the basis of such affective knowledge—which might well be argued to form the basis for the countertransference (in its modern usage). These findings offer, therefore, the possibility of that most unlikely of things: a neurobiological explanation of intuition.

This concludes our brief review of the neuropsychology of memory from the viewpoint of the "inner world." There is a great deal more that could have been said, as memory is one of the most popular research topics in modern neuroscience. There is, of course, also a lot more that could be said about the implications of these research findings for depth psychology. Nevertheless, we have managed to cover most of what is fundamentally important to our theme. We are now in a good position to tackle our next topic: dreams and hallucinations.

CHAPTER 6

DREAMS AND HALLUCINATIONS

Dreams are the primary focus of this chapter. They are hallucinations that we *all* experience—hallucinations that have been regarded by many as a "normal" form of psychosis. Freud was especially interested in dreams because he believed that, if he could understand their mechanism, he would be able to comprehend something fundamental about mental illness.[1] In the later sections of this chapter, other forms of hallucination and delusion, primarily in schizophrenia, are discussed. This chapter focuses on many of the same brain structures that were covered in the previous three chapters on consciousness, emotion, and memory. This is because the brain mechanisms of dreaming (perhaps not surprisingly) overlap a great deal with those of consciousness, emotion, and memory.

DIFFICULTIES INVESTIGATING DREAMING

Dreams are notoriously difficult to investigate scientifically. This chapter is therefore also concerned with methodological questions regarding *how* brain mechanisms of dreaming have been

[1] Many other physicians and scientists shared this view; see Gottesman (1999, pp. 470, 500) for examples.

investigated. Attention is drawn to the dangers of using inappropriate methods to investigate complex psychological states, and to the advantages of using more than one scientific method to study a difficult and elusive subject. In the past, one of the failings of psychoanalysis was its overreliance, despite the great complexity of its subject matter, on a single method for reaching its conclusions, but this has begun to change somewhat in recent years. Checking the findings of one method against those of another makes it possible to minimize the bias associated with a single method. Our review of the dreaming brain draws on findings from neurophysiological work on animals, sleep studies and functional-imaging studies in neurologically intact humans, and clinical and experimental investigations of patients with focal brain lesions.

REM SLEEP

Any discussion of the brain mechanisms of dreaming cannot begin before the phenomenon of **rapid-eye-movement** (REM) sleep has been introduced, because REM sleep has become widely known as "dreaming sleep." However, as we shall learn, it is a mistake to equate the two phenomena. Indeed, the conflation of REM sleep with dreaming is one of the most substantial errors that has arisen from methodological impropriety in this field.

When the REM state was discovered in the 1950s, the scientists involved (Aserinsky & Kleitman, 1953; Dement & Kleitman, 1957) immediately suspected that it might be the physiological correlate of dreaming. This was because the REM state involves a period of physiological arousal in the context of otherwise quiescent sleep, just as the dream state involves conscious mental activity in the context of otherwise unconscious sleep. During

REM, it is not only the eyes that are active. An electroencephalogram (EEG)—which provides a measure of the electrical activity in the brain—made during REM would suggest that although you are sleeping, your brain is in a state of heightened activation akin to full wakefulness. There is also activation of other bodily systems. You begin to breathe differently, your heart rate increases, and your genitals (in both males and females) become engorged.[2] One is thus highly excited in several ways during REM sleep. By contrast, however, skeletal muscle tone *drops* dramatically (with the exception of the musculature controlling eye movements). This effectively paralyzes the sleeper, and it apparently prevents him or her from acting out dreams. This cycle appears more or less every 90 minutes in humans, so that we spend some 25% of our sleeping hours in the REM state.

The easiest and most obvious way to test the hypothesis that the REM state is the physiological equivalent of dreaming is to wake people up during both REM and non-REM sleep and compare the frequency of dream reports found in the different awakenings. The first time this hypothesis was tested, it was immediately obvious that many more dream reports are obtained from REM than non-REM (NREM) awakenings. Today, 50-odd years after this issue was originally investigated, there remains some controversy about the exact percentages. The strongest claim is that 90–95% of awakenings from REM sleep produce dream reports, whereas only 5–10% of awakenings from NREM sleep produce equivalent reports. Probably most authorities would agree on a conservative 80:20 (REM:NREM) dream-report ratio.

[2] In fact, penile erection during REM sleep is so reliable that it provides the basis for one of the most widely used investigations of male impotence. If one measures penile tumescence during sleep and the subject has erections during REM, it is likely that his impotence is of psychological origin.

Taking into account the fallibility of human memory in general, let alone memory for *dreams* (which are particularly difficult to recall), it would have been unreasonable for early investigators to expect to obtain a 100% dream recall rate from REM sleep awakenings, or a 0% rate from NREM awakenings. Under the circumstances, the (roughly) 80:20 ratio that *was* observed was therefore interpreted as a near-perfect correlation, and the hypothesis was taken as confirmed: it was concluded that REM sleep and dreaming were literally the *same thing*, considered from two different observational perspectives (see chapter 2). This equation provided an extremely valuable scientific foothold (although it later proved to be a slippery slope): by making the assumption that the REM state is synonymous with the dream state, scientists believed that they had in their grasp an objective measure of the presence or absence of dreaming. As a consequence, they could carry out objective experiments on perhaps the most subjective of all mental functions, the *psychological* study of which had, moreover, provided the theoretical bedrock for the whole discipline of psychoanalysis (which at that time totally dominated American psychiatry). The fact that not only humans but *all mammals* display the REM state made it possible for neuroscientists to go one step further: they could identify the *brain mechanisms* underlying the REM state (read: dream state) by means of animal experiments that were ethically unacceptable in humans. This is where the slippery slope began, for no matter how close the homologue may be between the REM state in humans and other mammals, we have no way of knowing whether the same applies to their *dreams*. The moment investigators switched from studying humans to other animals, the monitoring of their subjects' dreams (as such) was perforce abandoned.

The biological basis of REM sleep

The main thrust of the ensuing investigations took the form of lesion studies. A French neuroscientist, Michel Jouvet (1967), carried out the first key studies by performing a series of ablation experiments. Although REM sleep occurs in a remarkably wide variety of animals, cats were the main targets of this research—partly because their brains are so similar to ours, but no doubt also because they sleep for so much of the day! Jouvet made a series of slices through the neuraxis of the cat, starting at the highest level of the frontal lobes and moving progressively downward toward the brainstem. He then systematically investigated the effects on the sleep cycle. He wanted to ascertain the key lesion site that would obliterate REM sleep. To his amazement, he found that you could effectively detach the entire forebrain from the brainstem, and the REM state would still remain intact and would punctuate NREM sleep with the same monotonous regularity. The critical incision occurred only in the middle regions of the primitive brainstem, at the level of the *pons* (see chapter 1). Subsequent investigators confirmed that REM sleep can only be obliterated entirely by creating fairly large lesions in the pons (Figure 6.1) (Jones, 1979). In short, these studies demonstrated that, whatever REM sleep was, it was *causally*

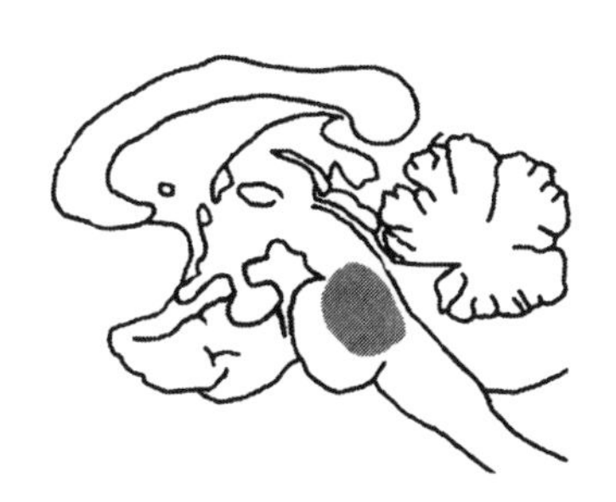

FIGURE 6.1
Lesion site to obliterate REM sleep

generated by structures in the pontine brainstem. The implications of this finding were enormous. Since the forebrain is the seat of all our higher mental functions (i.e., representational cognition; see chapters 1 and 2), the early investigators concluded that REM sleep (read: dreaming) is an entirely "mindless" activity. This raised serious questions for any psychological theory of the causation of dreams, not least among them being the Freudian theory that dreams are caused by wishful states of mind. The following quotation is from one of the most influential papers in the field:

> If we assume that the physiological substrate of consciousness is the forebrain, these facts completely eliminate any possible contribution of ideas (or their neural substrate) to the primary driving force of the dream process. [Hobson & McCarley, 1977, p. 1338]

Links between REM sleep, consciousness, and emotion

The role of the pons and other nearby brainstem structures in creating "core" consciousness (discussed in chapter 3) is not incompatible with the notion that dreams are "mindless." Nobody denied that dreaming is a *state of mind*, that you are *conscious* while you are dreaming. The same applies to the fact that many dreams are strongly *emotional* experiences. Although the role of the PAG in generating emotional states had not yet been discovered, the role of the reticular formation in generating consciousness *was* known at that time. But whether these links were understood or not was immaterial: the early investigators did not deny that dreams *took the form* of conscious, emotionally charged experiences; all they asserted was that the mental aspect of dreaming was not *causal* of the dreams. Dreams, they argued, are caused by something happening in the pons that switches on absolutely *automatically*, every 90 minutes or so, regardless of your state of mind. Since nearby brainstem structures were also known to regulate eye movements, heart rate, and breathing, it seemed perfectly obvious that REM/dreaming was just a basic physiological state. The biological reason for this pontine clockwork was (and remains) unknown, but it was confidently assumed that dreams were merely a byproduct (or *epiphenomenon*) of this causal physiological process.

Philosophically minded readers might have problems with this type of reasoning. You might well ask (as we did in chapter 2) whether it ever makes sense to claim that a physiological process *causes* a mental event, and vice versa, or whether it makes sense to claim that some neurophysiological events are mindless while others are not. From the standpoint of dual-aspect monism (see chapter 2), every neurophysiological event is simultaneously a mental event—albeit, ultimately, an unconscious one. However, although the early neuroscientific investigators of REM dreaming did not address such issues in any depth, they were able to claim that since the generation of REM is an automatic, preprogrammed process, its unconscious mental correlate is as "motivationally neutral" (Hobson & McCarley, 1977, p. 1338) as the brainstem mechanism that generates your heartbeat. This much seemed certain.

The neurochemistry of REM

By 1975, Hobson and McCarley had narrowed the search for the pontine "dream-state generator" (as they called it) to a set of precisely defined nuclei within the pons. In that year they published a famous paper in which they argued that the REM state is switched on and off by two groups of reciprocally interacting nuclei, one of which excretes a neurotransmitter that switches it on, and the other two neurotransmitters that switch it off

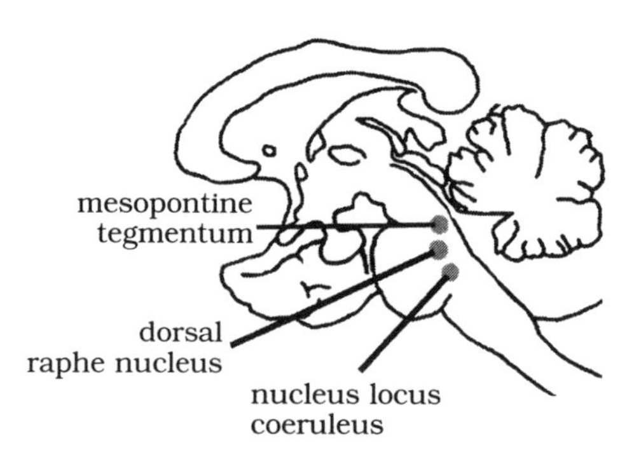

FIGURE 6.2
The "dream-state" generator

(Hobson, McCarley, & Wyzinki, 1975). Although they later changed their minds about some of the specifics of the anatomy, their argument was that the key neurons that switch REM sleep *on* lie in the **mesopontine tegmentum** (see Figure 6.2). These neurons fire rapidly shortly before the onset of REM, and they stay highly active throughout the REM phase. The neurotransmitter that these neurons produce is **acetylcholine** (see chapter 1). They are therefore described as *cholinergic* cells, and the REM state is considered to be a cholinergically driven phenomenon.

At the transition between REM and NREM sleep, two other sets of neurons, in the **dorsal raphe nucleus** and the **nucleus locus coeruleus,** start firing rapidly (Figure 6.2). The dorsal raphe produces the neurotransmitter **serotonin**; the nucleus locus ceroleus produces *norepinephrine.* When these nuclei become active, the cholinergic system simultaneously (in fact, *consequently*, due to reciprocal inhibition) switches off. This switches off the REM state, and the sleeper falls back into NREM sleep, with a mellowing surplus of serotonin and norepinephrine flowing around the brain. Some 90 minutes later, these two groups of nuclei reciprocally alter their function again—such that the levels of serotonin and norepinephrine drop, and acetylcholine increases—and the excited REM state reappears.

So, according to this model, acetylcholine switches the REM state on and switches NREM sleep off. Serotonin and norepinephrine switch NREM sleep on and switch the REM state off. For obvious reasons, this model was named *the reciprocal interaction model.* This physiological account of REM sleep is extremely compelling. Some 25 years after it was first proposed, it still completely dominates the field of REM sleep research. By 1975, then, some of the great mysteries surrounding sleep and dreaming appeared to have been resolved.[3]

Dreams are froth?

Two years after they proposed the reciprocal activation model, Hobson and McCarley published a second paper, which contained a second model—this time not a model of REM sleep but, rather, one of *dreaming* itself (Hobson & McCarley, 1977). This seemed to be a legitimate extension of the first model, because REM sleep and dreaming were thought to be essentially the same thing. They called their second model the *activation synthesis model.* The *activation* aspect of the model argues (no surprise here) that dreaming is activated by cholinergic mechanisms in the pontine brainstem. As we have noted already, this activation—which actually *causes* dreaming—is thought to be "motiva-

[3] It was subsequently discovered that there are interesting periodic changes in our mental state during *waking* life on a 90-minute cycle, which may well relate to the REM/NREM cycle during sleep. However, the effects are far more dramatic during sleep, in part because there is an enormous input of sensory material from external reality during waking life. During sleep (when the other inputs are removed), these intrinsic oscillatory mechanisms appear to dominate.

tionally neutral." The *synthesis* aspect of the model argues that the forebrain, thus activated, lamely attempts to piece together (or synthesize) the meaningless conscious representations (memory images, thoughts, and feelings) that are randomly stimulated from below. Again, note that the forebrain's contribution to the process is secondary to a primary, brainstem-driven events—hence the notion that the dreams themselves are epiphenomenal to the REM state. From the forebrain's point of view, images are being activated during REM as if it were wide awake and experiencing something; it therefore does the only thing it can do, which is to string the images together into a self–object *episode*. In Hobson and McCarley's memorable phrase, the forebrain "does the best of a bad job" by trying to make a sensible experience during REM sleep out of the intrinsically "inchoate" images thrown up by the brainstem (Hobson & McCarley, 1977, p. 1346). Freud had a term for this sort of theory, which actually existed in a speculative form in 1900, prior to the development of modern neuroscience. The expression is "*Träume sind Schäume*," which literally translates as "dreams are froth" (Freud, 1900a, p. 133 [p. 138 in the original German]).

This phrase epitomized Hobson and McCarley's conception of dreams. Their work posed a clear threat to Freudian psychoanalysis, and Hobson wasted no time in pointing this out at the 1976 annual meeting of the American Psychiatric Association (APA). After Hobson's presentation, a vote was taken among the membership of the APA as to whether Freud's dream theory was still scientifically tenable, in the light of Hobson's findings. At that time, the APA was still dominated by members sympathetic to psychoanalysis. Nevertheless, the result of the vote went overwhelmingly against Freud—suggesting that this was the end of the road, scientifically speaking, for Freud's (1900a) account of the mechanism of dreams. Given that Freud saw dreams as the "royal road" to an understanding of the unconscious mind, this had serious implications for psychoanalysis in general. It is no exaggeration to suggest that the tide turned decisively against psychoanalysis in America at that fateful 1976 meeting.

The dreams of cats

But, as the attentive reader will recall, the activation synthesis theory embodied a critical methodological flaw. Hobson and McCarley's dream theory rested heavily on the *assumption* that dreaming and the REM state are synonymous. The finding that the REM state co-occurs with dreaming in humans, and the fact that the REM state also occurs in cats (and rats), led to a series of experiments on the brains of these lower mammals that sought to identify the brain mechanisms that generate REM sleep (read: dreaming). Having done so, and having demonstrated conclusively that only large pontine-brainstem lesions obliterate REM sleep, the next logical step would be to check whether these lesions also obliterate *dreams*. After all, it was largely the association between REM sleep and dreaming that made it so interesting in the first place. The problem was, of course, that it is impossible to ask a cat (or a rat) whether it is dreaming or not. Some cat lovers might believe that they *do* know when their beloved animals are dreaming, but even behaviorists know that it is dangerous to infer the content of an inner mental state from the form of an external behavior!

The fact is that a reliable method for checking the assumption that the REM state and dreaming are synonymous was available all along, by investigation of dreaming in those who can provide a subjective report. However, once neuroscientists had made the assumption that REM and dreaming were synonymous, it had become such a truism that nobody seemed even to think of checking it, and attention shifted naturally to animal research.

A reliable method for linking a psychological function with a brain structure is the *clinico-anatomical method*, which forms the basic methodological building-block of human neuropsychology (see chapter 2). This well-established tool was introduced to neuropsychology by Pierre Paul Broca in 1861. The *clinical* side of this method involves making an observation that a mental function is lost following a focal brain lesion. In Broca's famous case, discussed in chapter 2, it was language that was lost. The *anatomical* side of the clinico-anatomical method involves ascertaining the precise extent and location of the brain damage that caused the loss of the mental function in question. In Broca's time, investigators had to "wait" for their patients to die to make this kind of observation. Today, this research can be conducted with living human subjects, using brain-imaging technology.[4] The autopsy of Broca's patient, Leborgne, revealed an area of damage on the lower left-hand side of the frontal lobe. Broca concluded that this is the neurological substrate of the ability to speak—because when it is damaged, speech is lost.

This turned out to be a somewhat oversimplified conclusion. We now know that other parts of the brain participate in a complex functional system subserving speech and language (see chapter 2); however, even these other components of the neural substrate for language were identified using the clinico-anatomical method. From 1861 onwards, therefore, the guiding principle in neuropsychology has remained the same: to demonstrate that activity in a certain part of the brain is the neural correlate of a specific mental function, it is necessary to link damage to that region to a deficit of that function.[5] Jones (1979) demonstrated this, for REM sleep, in cats; subsequent sleep researchers confirmed that this clinico-anatomical correlation held good for humans too (in cases with naturally occurring lesions). The link between the pons and *REM sleep* is therefore clearly established,

[4] Computerized tomography (CT) and magnetic resonance imaging (MRI).

[5] Today it is possible to go further, and check the clinico-anatomical correlation by ensuring that exercising the function in question is linked with increased metabolic activity in that same region (using PET imaging and fMRI).

in humans and other animals. But it is only in humans that the link with *dreaming* could be established—or refuted.

REM and dreams are not synonymous

Astonishing as it may seem in retrospect, the equation "REM sleep = dreaming" was subjected to systematic clinico-anatomical scrutiny only forty years after the association between REM sleep and dreaming was discovered. And when it was, it was found to be seriously wanting. In a 1997 study, 6 patients who had sustained damage to the REM-generating regions of the pons were asked whether or not they were still dreaming, and their answer was a clear "*yes*." In contrast, more than 40 other patients with damage to specific parts of the *forebrain*, nowhere near the critical REM-generating structures, *did* experience a cessation of dreaming following their brain damage—but in these patients *the REM state was preserved* (Solms, 1997a; see also Solms, 2000a).

NON-REM DREAMS

The discovery that pontine-brainstem damage does not produce a loss of dreaming in humans led researchers belatedly to take account of previously neglected observations that seemed to point in the same direction (derived from other methods, more suited to human dream research than research on cats and rats). The main victim of this neglect was the work of David Foulkes (a Chicago psychologist) and his collaborators. Foulkes (1962) focused on NREM dreams, which, according to classical teaching, were supposed to be extremely rare. What he found was that by simply rephrasing the question that subjects are asked upon awakening in the sleep laboratory—saying to them

"What was passing through your mind?" rather than "Were you dreaming?"—subjects reported that they were experiencing complex mentation during NREM sleep on 50% of awakenings. However, the REM = dream theorists were quick to point out that dreaming is not the same as thinking.

Our attention therefore shifts to the 5–10% of occasions on which subjects report fully fledged dreams in NREM sleep. These dreams are no different from the dreams of REM sleep. Even Hobson (who has the most to lose from such findings) confirmed that these NREM dreams are "indistinguishable by any criterion" from REM dreams (Hobson, 1988, p. 143). In fact, some REM = dreaming theorists believed that these *were* actually REM dreams that were misattributed to NREM sleep due to the aforementioned vagaries of human memory.[6] Foulkes (1962) showed that this assumption was wrong. He observed that you are most likely to have dreams in NREM sleep *just after you have fallen asleep*, during what is known as the *sleep-onset* phase of NREM sleep (more technically known as Descending Stages 1 & 2). On awakenings from these first few minutes after falling asleep, subjects report dreams some 70% of the time. Most people do not remember these dreams when they wake up in the morning, for obvious reasons, but we have all had the experience of dozing off briefly and then waking up (often with a start) from a dream. These dreams occur before you have entered your first REM period (in fact, roughly 90 minutes before). The 70% of NREM

[6]A similar claim has recently been advanced by Tore Nielsen (2000). Nielsen suggests that although these dreams occur during NREM sleep, as defined by the standard physiological criteria endorsed by the field for over 30 years (Rechtschaffen & Kales, 1968), they are probably generated by intrusions of REM physiology into the NREM state. NREM dreams, according to Nielsen, are therefore actually "covert" REM dreams. Hobson so enthusiastically grasped Nielsen's lifeline to his theory that he actually went so far as to assert (in Hobson, Pace-Schott, & Stickgold, 2000) that "all sleep is REM sleep, more or less"!

dream reports that are obtained from the sleep-onset phase therefore cannot be misremembered REM dreams.

Antrobus and his colleagues made a related observation, at the opposite end of the sleep cycle (Kondo, Antrobus, & Fein, 1989). They demonstrated that the closer you get toward awakening in the morning—that is, at the end of a night's sleep, after the last REM phase (more technically: in the rising morning phase of the diurnal rhythm)—the more likely you are to obtain a REM-like NREM dream report. This is called the "late-morning effect." The implications of this finding are similar to those for sleep-onset dreaming: the *farther away* from the last REM period you get, the more likely you are to have an NREM dream. In the classical "reciprocal interaction model," these transitional phases between wakefulness and sleep (sleep onset and the late morning) were described—physiologically speaking—as *maximally distinct from the REM state*: they were characterized by very high levels of norepinephrine and serotonin and very low levels of acetylcholine. Clearly, then, dreaming is *not* causally dependent on the unique physiological characteristics of the REM state. However, most NREM dreams do share *another* crucial characteristic with the REM state, which probably casts important light on their true causal physiology. We shall mention this characteristic in a moment.

Before doing so, it is worth asking in passing why the findings contradicting the REM = dreaming doctrine were neglected for so long. The answer may have something to do with the difference between "brain" observations (concerning the state of a piece of physical tissue), and "mind" observations (concerning the contents of subjective reports). The reaction of the scientific community to findings in dream research, and perhaps in other aspects of neuroscience, have often been distorted by the fact that we are more prepared to accept evidence derived from precisely measurable physiological and anatomical variables than from the complicated field of clinical and subjective reports. Understandable as this bias may be, the example of dream research shows that it is essential for modern neuroscientists to take serious account of the data derived from *both* observational perspectives in the mind–body equation.

DREAMS AND AROUSAL

The feature that most NREM and REM dreams have in common is *arousal.* This term is not implied in the narrow sexual sense, but with reference to levels of brain activation. Shortly after you fall asleep, your brain is still relatively aroused, as you begin the gradual decline from full wakefulness into sleep.[7] As mentioned earlier, the REM state is characterized, perhaps above all, by sustained periods of (cholinergic) brain activation interrupting an otherwise quiescent sleep state.[8] The rising morning phase, too, is characterized (indeed, defined) by relative arousal—albeit hormonally rather than cholinergically mediated. The three periods of sleep during which you are most likely to experience a dream, therefore, are characterized not by the unique physiology of the REM state (which characterizes only one of the three periods) but by *various types* of arousal. This suggests that a certain *amount* rather than a certain *type* of arousal is a necessary precondition for dreaming.

In the activation synthesis theory, the arousal that accompanies dreams was not only thought always to be of the same type (i.e., *cholinergic* arousal), it was also thought always to arise from the same place (namely, the *brainstem*). If this were true, it might still be possible to claim that dreams are "mindless" and

---

[7]The contribution that these remnants of wakefulness make to dreaming may be one source of what Freud (1900a) called the "day residues" in dreams.

[8]This striking coexistence of heightened brain activation with ongoing sleep led the early investigators of what later came to be known as "REM sleep" to call it "paradoxical sleep."

"motivationally neutral." But, in fact, good evidence exists that suggests that dreams *can* be causally generated by *forebrain* mechanisms.

DREAMS AND EPILEPSY

There is a form of epilepsy that involves *partial* seizures that are entirely localized to the limbic regions of the forebrain. Partial seizures occur when the abnormal neuronal activity that causes a seizure does not spread to the rest of the brain (which normally causes seizures to *generalize* into the familiar form known as "tonic clonic convulsion"). Partial seizures reflect their localization: if epileptiform neuronal firing occurs in the visual cortex of the right occipital lobe, the seizure takes the form of flashes of light (or "phosphenes") in the left visual field; if the abnormal activity is in the left motor cortex, the seizure takes the form of twitches in the right arm or leg. Similarly, when epileptiform brain activity is localized to the *limbic* parts of the forebrain, which subserve emotional and memory functions (e.g., the amygdala and hippocampus; see chapters 4 and 5), the resultant seizure takes the form of a *complex mental experience* (e.g., a reminiscence accompanied by a strong feeling of emotion). This limbic form of partial seizure is called *complex*-partial to distinguish it from the elementary sensations and movements that are characteristic of the *simple*-partial seizures just described.

Seizures occur quite frequently during sleep, and typically during the NREM phases—which are characterized by rhythmic, slow waves of electrical activity of a kind that are apt to set off seizures in predisposed brains. These seizures assume various forms (depending on the location and extent of the epileptogenic focus), but not infrequently they take the form of complex-partial seizures. This implies (by definition) that the abnormal brain activity causing the seizure is *wholly confined to the limbic regions of the forebrain.* Specifically, the seizure focus does not spread to the core brainstem structures that regulate the sleep cycle (if it did, the resultant seizure would be neither complex nor partial). It is therefore of considerable interest to observe that these NREM seizures are frequently accompanied by dreams. In fact, they are typically associated with highly distinctive dreams and take the form of recurring nightmares (which reflects the involvement of limbic emotional and memory mechanisms). Given what we know about the underlying physiology of these dreams (which are, in fact, seizures—unequivocally *caused* by focal activation of specific limbic forebrain structures during NREM sleep), one may confidently conclude that the arousal mechanism that triggers dreams is not *necessarily* located in the brainstem at all. Dreaming, it seems, can be triggered by arousal of *any type* arising from *any place*—including the emotion- and memory-generating structures of the limbic forebrain. This casts further serious doubt on the assertions of the old REM/brainstem dream theorists who claimed that the activation of ideas, memories, and emotions cannot, to quote Hobson and McCarley (1977) again, be "the primary driving force of the dream process."

WHAT IS THE "PRIMARY DRIVING FORCE" OF THE DREAM PROCESS?

If it is no longer tenable to assert that the pontine brainstem contains the primary causal generator of dreaming, then what *is* the primary driving force behind dreams? We said earlier that clinico-anatomical studies revealed that lesions of the pons did *not* cause cessation of dreaming (evidence *against* an exclusive causal role for the pons), but we also said that lesions in two forebrain regions *did* have that effect. Do *these* regions perhaps contain the long-sought "dream-state generator"?

The first of these regions is the transitional zone between the occipital, temporal, and parietal cortex, at the back of the forebrain, in the very hub of the functional unit for receiving, analyzing, and storing information (see chapter 1). Lesions in this area (on *either* side of the brain) produce a total cessation of dreaming (the precise location of these lesions, though, is still uncertain: see Yu, in press).

The other region with this property is the limbic white matter of the ventromesial quadrant of the frontal lobes. Damage to this area of the brain (on *both* sides simultaneously) also produces a total cessation of dreaming. Damage to other parts of the brain causes other characteristic *changes* in dreaming (e.g., increased frequency of dreaming, increased nightmares, defective visual dream imagery). This suggests that these regions, too, form part of the complex "functional system" that generates dreams (see chapter 2). The parts of the brain in question include the entire *limbic system* (including all the "limbic" components of the frontal and temporal lobes, but excluding most of their "higher cognitive" components), as well as most of the *visual system* (excluding the visual "projection" cortex). However it seems likely that one of the two structures that are *essential* for the generation of dreams (i.e., either the occipito-temporo-parietal junction or limbic frontal white matter) embody the "primary driving force" behind dreaming.

FUNCTIONAL-IMAGING FINDINGS

We said earlier that clinico-anatomical findings nowadays are typically checked against functional-imaging findings for accuracy. This is in keeping with the view that scientific conclusions regarding something as complicated and experimentally elusive as human mental life should—wherever possible—be confirmed by multiple, converging methods of investigation before they can be accepted with confidence.

Through functional brain imaging, it is possible to obtain a graphic representation of the brain of a healthy living subject and to observe where neural activity is greatest during certain mental states. In the last few years, this procedure has been applied to sleep and dreaming by a number of pioneering investigators. The definitive studies in this regard were published by Alan Braun from the National Institutes of Health in Washington, DC, who, together with colleagues (Braun et al., 1997, 1998), used PET to investigate what the brain looks like during REM sleep—the time when one is *most likely* to be dreaming.[9]

During such investigation of the state of the brain during REM sleep, one is probably imaging two different states simultaneously: the REM state and dreaming. There is an 80% chance that dreaming will occur during REM, so the average of the data spread across several REM phases will almost certainly also capture the dream state. (With PET imaging it is always necessary, for technical reasons, to study the *average* picture.) The picture that emerges is therefore a combination of the dreaming and the REMing brain. Not surprisingly, then, Braun found that the pontine brainstem mechanisms that switch on the REM state were highly active during REM sleep. More interesting is what *else* he found.

The activation synthesis theory would have predicted that the brainstem activation of REM should globally activate the entire forebrain—thereby generating the random sensory, motor, emotional, memory, and thought images that comprise the supposed

---

[9] Other investigators have carried out similar studies, and all have produced findings compatible with Braun's. Due to technological constraints, there have, as yet, been no functional-imaging studies of the brain during sleep onset, or the late morning, when dreaming is dissociated from the REM state. However, these constraints will soon be removed when fMRI technology is applied to dreaming sleep.

"froth" of dreams. This is not what Braun found. Instead, he observed that only highly specific parts of the forebrain were activated during REM dreaming, while other parts were completely inactive. This is evidence of a striking pattern of dissociation between the levels of activation of various parts of the forebrain during sleep, suggesting that dreams are constructed by highly specific forebrain mechanisms. Moreover, the parts of the forebrain that Braun found were most active during dreaming were precisely the parts that obliterated or otherwise altered dreaming when they were damaged by brain lesions—and, conversely, the least active parts were the parts that had no effect on dreaming when damaged (Solms, 1997a). Braun therefore observed the very same pattern of dissociation that the lesion studies had found: the parts of the forebrain involved in the construction of dreams are the entire *limbic system* (including all the "limbic" components of the frontal and temporal lobes but excluding their "higher cognitive" components) as well as most of the *visual system* (excluding visual "projection" cortex). This implies, among other things, that the brain mechanisms of dreams are the same as those for the basic emotions discussed in chapter 4.

THE DREAMING BRAIN AND THE EMOTIONAL BRAIN

Let us quickly review the basic-emotion command systems. There is the SEEKING system, which runs from the transitional area between brainstem and forebrain to the limbic components of the frontal and temporal lobes (Figure 6.3). It is a nonspecific motivational system engaged in looking for something to satisfy needs. The SEEKING system is linked to the pleasure/lust subsystem, involving nearby basal forebrain nuclei—especially the nucleus accumbens. The RAGE system involves the amygdala (in the limbic temporal lobe) and the hypothalamus and upper

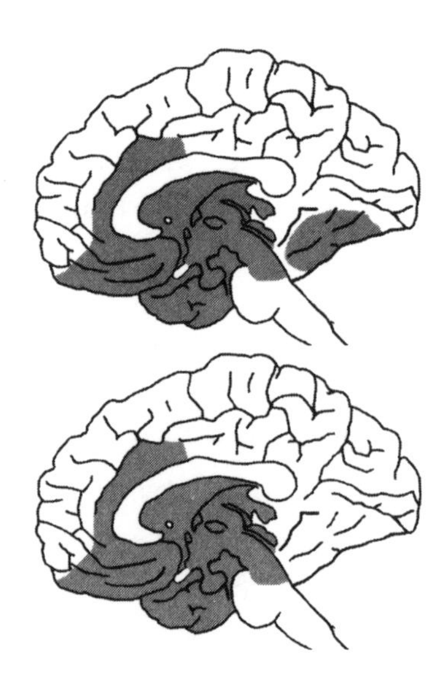

FIGURE 6.3
Top: The dreaming brain. *Bottom:* The emotional brain

brainstem structures. The FEAR system follows a very similar course. The PANIC system courses down from the anterior cingulate gyrus (in the limbic frontal lobe) to the same upper-brainstem structures. All of these emotional systems (together with the hippocampus, which subserves episodic memory, and parts of the visual system) are highly active during REM dreaming. But which of them provides the "primary driving force" of dreaming?

THE PRIMARY DRIVING FORCE BEHIND DREAMING, REVISITED

We said earlier that a certain degree of *arousal* was a necessary precondition for dreaming. We also said that two *forebrain* structures are essential for the generation of dreams (i.e., the occipito-temporo-parietal junction and the limbic frontal white matter). One of these two regions, we said, therefore probably contains the "primary driving force" behind dreaming.

While arousal is a *necessary* precondition for dreaming, it is not a *sufficient* condition to produce dreaming. We know this to be the case from the observation that patients with damage in the occipito-temporo-parietal junction or the limbic frontal white matter cannot dream, no matter how aroused they may become during sleep (even in the REM state). The necessary and sufficient conditions for dreaming are (1) forebrain arousal and (2) integrity of the occipito-temporo-parietal junction and limbic frontal white matter. So, which of the latter two structures provides the primary driving force? One way of addressing this question is to consider what else these latter two structures are known to do.

The occipito-temporo-parietal junction is heavily implicated in the generation of visuospatial imagery (see Kosslyn, 1994), and it is therefore no surprise that it should be implicated in dreaming—which is, after all, a special type of visuospatial imagery. The limbic frontal white matter, on the other hand, has no known function that is obviously implicated in dreaming, though a link to it may lie in a formerly common surgical procedure.

FRONTAL LOBOTOMY AND DREAMING

From the 1940s until the 1960s the dramatic procedure of frontal lobotomy (surgical disconnection of the prefrontal lobes) was performed on thousands of patients for the treatment of serious mental illness, especially schizophrenia.[10] In the early

[10] The procedure was developed in 1935 by Egas Moniz and Pedro Almeida Lima but only became widely used in the late 1940s. The term "lobotomy" was later replaced by "leucotomy," when the surgical target was reduced from the whole lobe to only a part of the underlying white matter.

days, this procedure involved a near-total disconnection of the prefrontal lobes from the rest of the brain. This certainly seemed to improve psychotic symptoms—especially the so-called positive symptoms of schizophrenia, such as delusions and hallucinations—but it also produced a range of debilitating side-effects. The most commonly reported adverse effects were inertia and apathy, intellectual decline, personality change, and postoperative epilepsy. The patients who underwent these operations lost not only their psychotic symptoms but also a great deal of what it means to be human.

For these reasons, some of the surgeons involved modified the procedure. They developed a more limited operation that damaged a far smaller region of the brain, with the intention of achieving the same therapeutic benefit but without the side-effects. Several different approaches involving different parts of the frontal lobe were attempted. They finally settled on the white matter underlying the ventromesial quadrant of the frontal lobes (for a review see Walsh, 1985, pp. 158–168). This modified procedure was called *ventromesial leucotomy* and involved using a custom-designed surgical "leucotome" to create bilateral lesions beneath the ventromesial surfaces of the frontal lobes (Figure 6.4).

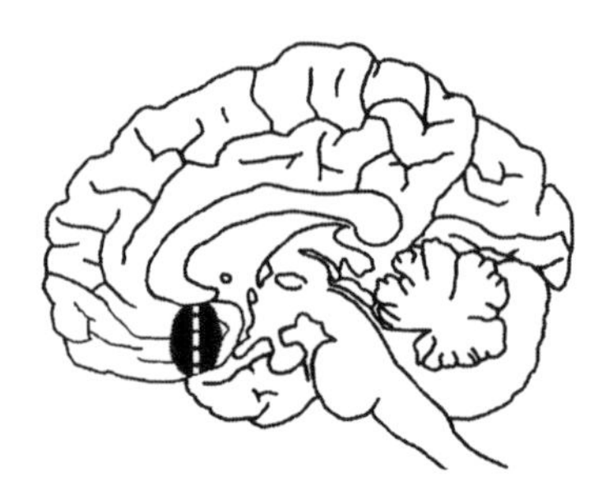

FIGURE 6.4
Modified frontal leucotomy

The area targeted by this modified procedure is exactly the one that the lesion studies mentioned above discovered was essential for the preservation of dreaming. In fact, we should say *re*discovered, for, incredible as it may seem, the practitioners of frontal leucotomy had already observed in the 1950s that the operation resulted in a cessation of dreaming in the vast majority of patients (for a review see Solms, 1997a, pp. 45–53). Psychiatrists knew this long ago, but, after the operation fell out of use, the knowledge was never transferred to the neuroscience literature. One psychiatrist even went so far as to observe that preservation of dreaming after the operation was a poor prognostic sign—continued dreaming seemed to indicate that the psychosis had not been successfully treated (Schindler, 1953). So, we may conclude that whatever it is that generates positive psychotic symptoms might well be the "primary driving force" behind dreams. As already noted, many psychiatrists have observed that dreams and psychotic illness must somewhere share a common mechanism.

DRUGS AND DREAMING

There are a number of reasons why the psychosurgical treatment of schizophrenia fell out of favor, including ethical considerations. However, it is widely accepted that the most important reason for the shift was the development of *pharmacological agents* that were just as successful at controlling positive psychotic symptoms (if not more so), but with fewer side-effects. These drugs are the "major tranquilizers," also referred to as "neuroleptics" or "antipsychotics." Psychiatrists still employ versions of these drugs today for the treatment of schizophrenia. All of these agents have one core feature in common—they block *dopamine* transmission, and mesocortical–mesolimbic dopamine transmission in particular (see chapter 1; for reviews see Lickey & Gordon, 1997, or Snyder, 1999). The mesocortical–mesolimbic pathways course through exactly the area of white matter that was targeted in ventromesial leucotomy (see Figure 4.3). For this reason, some neurobiologists have irreverently remarked that antipsychotic drugs function as "chemical leucotomies" (Panksepp, 1985, p. 273).

In chapter 4 we pointed out that one of the basic-emotion command systems is embedded in precisely these same pathways. This is the dopaminergic SEEKING system. Antipsychotic medications therefore block activity in this system, just as the old ventromesial leucotomy procedure did.[11] This blocking treats the positive symptoms of schizophrenia because, for reasons not well understood, overactivation of the SEEKING system seems to generate the positive symptoms of schizophrenia. This association is demonstrated, among other things, by the fact that pharmacological *stimulation* of this system can artificially produce psychotic symptoms in psychiatrically normal subjects. *Cocaine* and the *amphetamines* are two other classes of pharmacological agent that act on this dopamine system. Lower doses of both of these drugs produce a great boost of energy and increased interest in objects in the world. This is consistent with increased activation of the SEEKING system. In higher doses, a "stimulant psychosis" (Snyder, 1999, pp. 138–140) is initiated. Long before dosage levels produce psychosis, however, users develop the feeling that some events in the world have a "special signifi-

[11]The system is not typically described by psychiatrists as the SEEKING system—this term is employed only by those working on the neurobiology of emotion. However, both literatures know this set of pathways as the mesocortical–mesolimbic ascending dopamine system. Psychiatrists often describe this set of pathways as the D2 (or 2nd dopamine) system. Thus, different disciplines have developed separate terminologies to describe the same neuroanatomical and neurochemical systems.

cance" for them, and they exhibit a degree of suspiciousness about the behavior of others. In the more extreme state, patients almost invariably become paranoid, and they sometimes suffer auditory hallucinations. Such stimulant psychoses can be rapidly and effectively treated by the administration of the antipsychotic medications usually given to schizophrenics.

The same thing can happen with the administration of dopamine agonists (stimulants) in the treatment of Parkinson's disease (the drug *L-dopa*, for example, is notorious for inducing psychotic symptoms). On this basis, Ernest Hartmann conducted a study that might be considered a direct test of the hypothesis that the mesocortical–mesolimbic dopamine (SEEKING) system is the "primary driving force" behind dreams (Hartmann et al., 1980). He administered either L-dopa or a placebo to neurologically and psychiatrically normal subjects, shortly after the first REM period. The effect was immediate and dramatic. The subjects who received the L-dopa experienced a massive increase in the frequency, vivacity, emotional intensity, and bizarreness of dreaming. The frequency, density, and length of their REM periods was, by contrast, completely unchanged. This provides further evidence for the dissociation between dreaming and REM sleep discussed above and suggests that the dopaminergic SEEKING system might well be the "primary driving force" we have been looking for.[12]

In summary, when the SEEKING system is damaged, patients lose interest in objects in the world, dreaming ceases, and positive psychotic symptoms (hallucinations and delusions) decrease. Conversely, when the system is stimulated, energy levels increase, dreaming increases, and psychosis may ensue. There

[12] This conclusion is controversial and is still hotly contested by Hobson and his school. See Pace-Schott et al. (in press) for an overview of all the arguments for and against this view.

is therefore a clear series of links between dreaming, psychosis, and the operation of the SEEKING system.[13] In Hobson's original argument against the Freudian dream theory, he said: "these facts completely eliminate any possible contribution of ideas (or their neural substrate) to the primary driving force of the dream process," and he argued that the real driving force behind dreams was "motivationally neutral" (Hobson & McCarley, 1977, p. 1338). In the light of the present-day neuroscientific evidence, it seems quite inappropriate to claim that dreams are not caused by "ideas" and that they are instigated by a "motivationally neutral" process. On the contrary, dreaming and motivated ideas (akin, perhaps, to Freudian "wishes") appear to be inextricably interlinked.

VISUAL AREAS INVOLVED IN DREAMING

We have said that there is a second forebrain area that appears to be critical for dreaming, but it seems less likely that this area is the primary *driving force* behind dreams. The precise role that the occipito-temporo-parietal junction plays in the dream process is not entirely clear. It may well be that lesions to these sites produce loss of dreaming because of the role of these sites in mental imagery. If the patient loses the ability to generate a mental image, then inability to dream seems a logical consequence. If this argument is valid, then the effects of this second

[13]In this context, it is of some interest that Freud (1924b [1923], 1940a [1938]) believed that psychotic states resulted from an overwhelming of the ego by the libidinal (appetitive) drives (i.e., by the system that motivates our interest in objects in the world). Freud's position is therefore quite consistent with the fact that some aspects of psychosis (no less than dreams) appear to flow from an overactivation of the SEEKING system. A full discussion of this interesting possibility is beyond the scope of this book.

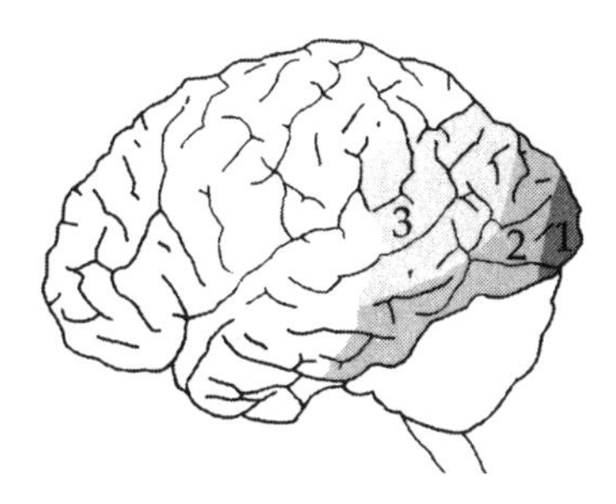

FIGURE 6.5
The three visual zones

lesion site are of less theoretical interest than is the important issue of the role of motivation in dreams.

A more significant finding is the *isolated loss* of visual dream imagery (or *aspects* of visual dream imagery, such as color or movement) after damage to the brain's visual areas. This suggests something about the "direction of flow" of information in dreams. The visual regions of the brain can be regarded as involving three hierarchically organized zones (see Figure 6.5):

1. The first is an area more or less directly connected to the retina, known as the primary visual cortex, lying at the back of the occipital lobes. This region is the "input" end of the system.
2. Next to this zone lies the "middle" part of the system, which is dedicated to a range of specialized visual-processing tasks. Color and motion processing, object recognition, and so on appear to take place here.
3. Finally, the zone in front of this is the highest level of the visual system, running the most abstract aspects of visual processing, which are also dependent on several other sensory modalities. This zone is involved in arithmetic, writing, constructional operations, and spatial attention. It represents the "output" end of the normal perceptual system.

Damage to Zone 1—the primary visual area—causes cortical blindness. Visual experience ceases, because the "input" end of the system is disrupted. Damage to Zone 2 causes more complex disorders of visual processing. These patients lose the ability to perceive color or movement, for example, or they lose the ability to recognize specific objects or faces. Damage to Zone 3—the occipito-temporo-parietal junction—does not affect visual perception *per se* but, rather, causes more abstract disorders that transcend concrete perception: acalculia (inability to calculate), agraphia (inability to write), constructional apraxia (inability to construct), and hemispatial neglect (inability to attend to one side of space).

In *dreaming*, however, this hierarchy seems to be reversed. Damage to the primary visual cortex, Zone 1, has (perhaps surprisingly) no effect on dreaming at all. Although these patients cannot see in waking life, they see perfectly well in their dreams. It seems that this aspect of the system is no longer the "input" end. Damage to the middle zone of the system, Zone 2, causes exactly the same deficits in dreams as it does in waking perception: these patients continue to dream in various sense modalities, especially somatosensory and auditory, but their visual dream imagery is deficient in specific respects. For example, they no longer dream in color, or they dream in static images (loss of visual movement), or they cannot recognize any of the faces in their dreams. Damage to the higher zone in the occipito-temporo-parietal junction, Zone 3, on the other hand, produces complete loss of dreaming. Recall that, in normal waking life, damage here does not affect perception *per se*, only higher visual cognition. That is, damage to the part of the system that is the "output" end in normal waking life seems to function as the "input" end during dreaming.

This "inverted" relationship has been proposed in the cognitive neuropsychological literature as a suitable model for the organization of the imagery system in waking cognition.[14] It seems to apply equally well to the case of dreaming. Freud called this mode of organization "regression" and wrote that "in dreams the fabric of thought is resolved into its raw material" (Freud, 1900a, p. 543).

A SUMMARY OF THE NEUROBIOLOGY OF DREAMS

The following description of the mechanism of dreaming integrates the evidence reviewed above. Some speculation is required, however, to fill in the gaps, although this will no doubt be less of a problem in the near future, as research in this exciting area is proceeding apace.

First and foremost, dreaming depends on a critical degree of activation of the basic mechanisms of core consciousness. If this inner source of consciousness is not aroused, then you cannot have a dream. It does not seem to matter what the trigger of the arousal is. Frequently, it is simply the residue of waking thoughts, as you drift off to sleep. The most reliable trigger is the REM state, which provides a sustained source of activation at regular intervals throughout sleep. As you begin to wake up, hormonal mechanisms gradually activate the forebrain. All these triggers activate (or "prime") consciousness, which is a necessary precondition for dreaming, but is not dreaming itself.

The activation of the motivationally charged SEEKING system, which drives our appetitive interest in the object world, appears to begin the dream process proper. It is probably accurate to say that *an arousal stimulus only triggers dreaming proper if it attracts appetitive interest.* When this happens, the

[14] Kosslyn (1994) calls it "backward projection." See also Zeki (1993).

subjective feeling is something along the lines of: "What could this be? I want to know more about this."

Activation of the SEEKING system during sleep is commonly, but not exclusively, triggered by the REM state. A thought process occurring during any stage of sleep can presumably also activate the SEEKING system. This thought process could be linked to an episodic memory from the previous day, or even just to a feeling. If the memory or feeling activates the interest of the SEEKING system, this would be enough to begin the dream process. This explains the observation that although *most* dreams occur at sleep onset, or during REM sleep, or just before waking, it *is possible* to have a dream at almost any time during the night—even during deep ("Stage 4") sleep. Recall, in this context, that these NREM dreams are indistinguishable from REM dreams.

When you sleep, you cannot go about exploring or seeking what you are motivationally interested in. This sort of behavior is not compatible with sleep, and it is probably for this reason that we dream. It seems a reasonable hypothesis that *the dream occurs instead of a motivated action.* That is, instead of doing something in the real world, you have a dream. The frontal lobes (the "action" end of the brain; see chapter 1) is normally a central "scene of action" in waking cognitive activity. However, this system is dormant (i.e., inhibited or underactivated) during dreaming sleep. The "scene of action" of cognitive activity therefore shifts to the posterior forebrain, with activation of the parietal, temporal, and occipital lobes. This is experienced as imaginative perception and cognition—which, however, differs from waking thought in that it is unconstrained by frontal inhibition. In the absence of the ability of the frontal lobes to program, regulate, and verify our cognition, affect, and perception, subjective experience becomes bizarre, delusional, and hallucinated.

In our dreams, the focus of motivated cognition is therefore removed from our goal-directed action systems and shifts toward the perceptual systems—especially the visuospatial component. The functional anatomy of dreaming is therefore almost identical to that of schizophrenic psychosis, as revealed by functional-imaging studies. One substantial difference is that in schizophrenia it is primarily the *audioverbal* component of the perceptual systems that is activated, rather than the visuospatial. The basis of this difference is unknown.

DREAMS AS THE GUARDIANS OF SLEEP

In addition to claiming that dreams were motivated by wishes, Freud famously argued that they "serve the purpose of prolonging sleep instead of waking up. Dreams are the guardians of sleep, and not its disturbers" (Freud, 1900a, p. 223). This means that the sleeper is "protected" from the disturbing influence of motivational urges that emerge during sleep. This hypothesis seems reasonable in the light of all we have learned above. But reasonable hypotheses are frequently proved wrong, and therefore untested hypotheses have limited value in science. One of the criticisms most frequently thrown at psychoanalysis is that its core hypotheses are untestable. It is here that the advantages of interdisciplinary collaboration become apparent: now that we know that certain unfortunate individuals with damage to specific parts of their brains are unable to dream, the sleep-protection hypothesis can easily be tested. Nondreaming neurological patients should experience sleep that is more disturbed than (say) that of neurological patients with equivalent degrees of brain damage in whom dreaming persists.

This critical test will soon be performed. It has only been possible so far to gather some preliminary data on the question, by simply asking patients who do not dream whether the quality of their sleep is unchanged, is better, or is worse than before their neurological problems began. The data collected so far (on a sample of 361 patients) supports Freud's sleep-protection theory at the required levels of statistical significance (Solms, 1995, p. 63). However, a sleep-laboratory study (instead of a bedside assessment) is required before this issue can be addressed with confidence.

DREAM CENSORSHIP

People who misunderstand Freud's theory of dream censorship (Braun, 1999; Hobson, 1999) mistakenly believe that the theory predicts that the (inhibitory) frontal lobes should be more rather than less active during dreams than in waking life (which is what they are). But Freud's dream theory states only that the "censorship" function of the executive ego is *not completely inactive* during sleep, not that it is *more active* during sleep than in waking life. In fact, according to Freud's theory, it is the *weakened* state of the inhibitory systems of the mind that makes our instinctual drives so unruly during sleep, and it is this that causes us to think and do in our dreams things that would be inconceivable in our waking lives. The theory therefore predicts exactly what functional brain imaging reveals—namely, that the inhibitory systems of the brain are *relatively* but not *completely* inactive during dreaming sleep (see Yu, 2000). However, this is far from proving Freud's censorship theory correct.

Freud's theory was designed to explain the differences between two components of the dream process. On the one hand, the manifest (or "explicit") content of the dream is often illogical and bizarre. On the other, the patient's associations to the individual elements of the dream suggest that the underlying, latent (or "implicit") content of the dream involved a motivational impulse that was not at all illogical or bizarre. In this regard, the neuroscientific evidence is compatible with Freud's model. Freud went on to question why the two levels of dream content differed so dramatically. His answer, of course, was the censorship. Here he may have been wrong. The apparent illogicality and bizarreness of dreams may be due to the inherently "regressive" nature of the dream process. The mere fact that the system is forced to function in the way that it does, where the executive systems of the frontal lobes cannot program, regulate, and verify the output of the posterior forebrain, may well produce the difference between the latent and manifest contents—with no need to introduce the additional function of censorship. The symbolic transformations to which Freud drew attention might, therefore, simply be the product of unconstrained parietal-lobe mechanisms operating in reverse, "resolving the fabric of thought into its raw material," as it were (Freud, 1900a, p. 543).

However, most observers would agree that the neuroscientific evidence does not yet have a decisive bearing on these important questions. The available evidence cannot tell us whether the distortions that appear to be introduced between latent and manifest dream thoughts are tendentiously motivated or not. For now, we shall have to rely on purely psychological techniques to assess the validity of this aspect of Freud's dream theory. Although multiple, converging lines of evidence are desirable in science, certain types of psychological question cannot be pertinently answered by neuroscientific methods.

In conclusion, modern neuroscience has come to understand a great deal about the biological basis of dreams, particularly the brain regions and attendant psychological processes that seem to be most central to the dreaming state. This knowledge is broadly consistent with Freud's psychoanalytic theory of dreams—although it would be inappropriate to say that his theory has been directly *proven*. The neural mechanisms of dreaming appear to overlap in several important respects with the neural mechanisms of certain core features of psychosis, especially the positive symptoms such as hallucinations. This confirms a long-standing hunch harbored by Freud (and many others) to the effect that understanding dreams might provide us with a key to understanding mental illness in general. Dreams truly do appear to be "the insanity of the normal man."

CHAPTER 7

GENETIC AND ENVIRONMENTAL INFLUENCES ON MENTAL DEVELOPMENT

The overwhelmingly vast topic of nature–nurture influences on the brain has the potential to include everything that neuroscience knows about the developmental sequence, in every psychological domain. The previous few chapters covered only *individual* mental functions, and they focused primarily on their organization in the mature, adult brain. This chapter broadens our focus considerably. We therefore want to emphasize at the outset that our goals in this chapter are very limited: to introduce some of the basic principles about genes and their workings and to discuss their implications for the broader theme of this book. The best way we could think of doing this was to begin by summarizing the main principles and then illustrating these with reference to *a single aspect of mental life*—thereby reverting to the structure of the previous four chapters. We decided to use *sexual difference* as our example, thereby enabling us to cover from a neuroscientific point of view (at least in part) another topic that has been a traditional stomping ground of psychoanalysis.

A PHOBIA ABOUT GENES

Many people appear to have something of a phobia about genes: an aversion to, or mistrust of, genetic "explanations" of behavior. This aversion seems to be based on the misconceived notion that genetic influences on behavior are something fixed and predetermined. This *would* be threatening—because if genetic influences are unalterable by experience, they are something that we can do nothing about. The true state of affairs is very different. Genetic and environmental influences on behavior are *absolutely inextricable*, and genetic influences are therefore anything but immutable. In fact, genes would be a terrible handicap if they were not accessible to environmental influences (consider, for example, the basic-emotion command systems discussed in chapter 4). Nature and nurture are in a dynamic interplay from the earliest moments of development.

TWO FUNCTIONS OF GENES

Genes are sequences of **deoxyribonucleic acid** (DNA) strung together in the famous *double-helix* structure to form **chromosomes**. Humans have 23 pairs of chromosomes. The gene sequences on these chromosomes have two functions, the conventional terms for which are *template* and *transcription* functions. Understanding the distinction between these two functions immediately explodes some popular myths about genes.

Many people know that genes replicate themselves. It is also well known that the genes of the male and the female are mixed at conception, and that the unfolding of this mixture produces the little genetic bundle of joy that we know as a baby. This ability to *replicate* is the *template* function of genes. It is unfortunate that most people imagine that this is all that genes do, because in fact nothing could be further from the truth. All of our genes are represented in every cell of our body, but their template function is—in a sense—restricted to the genes in the cells of the sperm and ovum. This leads to the important question of what all the genes in the rest of the body do, including the genes in the billions upon billions of cells that make up the nervous system. All of this goes under the heading of the *transcription* function of genes.

THE TRANSCRIPTION FUNCTION OF GENES

The transcription function of genes is closely bound up with what we call the "expression" of genes. The genetic codes (sequences of acids) making up the strands of DNA are designed to produce different proteins. In the simplest case, a particular protein thus produced will make your eyes blue or brown, and your hair black or red.

If all that a gene does is produce a protein, then how, you might ask, can there be a gene for schizophrenia, one for hyperactivity, one for criminality, one for alcoholism, and so on, as we so often hear in the mass media? How can a protein turn you into a criminal? Surely it is impossible to reduce such a complex psychological condition to the operation of a single protein? We would certainly agree that such arguments are dramatic oversimplifications. Genes create and modify various brain structures, and we have seen in earlier chapters that the neurobiology of, say, a given mental illness involves a wide range of brain regions, in ways that are undoubtedly overdetermined. It is also important to remember that genes do not work alone but, rather, in complex interactions with one another. To produce even one neural circuit—for example, one that performs the elementary perceptual function of registering light—a very complicated sequence of genetic events needs to take place. Therefore, even if it were possible for conditions such as schizophrenia and hyperactivity to be simply "programmed" by genetic events, the programs in questions would need to be extremely complex and involve a great many genes working in concert.

We have said that each of the cells of your body contains all of your genes. All them can therefore *potentially* produce a huge range of proteins. But in reality there is a division of labor between the different cells of your body. They do not all involve themselves in the full diversity of things that the human genome can produce. The genes in different cells produce proteins that represent only a small sample of their potential range—which is the same as to say that only a small percentage of the genes in any one cell are actually *expressed*. The difference between a liver cell and a brain cell arises from the fact that the genes that are expressed in them are different, resulting in the growth of different types of cell and ultimately (due to the clumping together of cells) different types of tissue. This is how the body comes to contain the great variety of organs and functions that it does.

The process of activating and expressing genes turns the **genotype** into the **phenotype**, transforming the *virtual* (or "potential") structure coded in your DNA into an *actual* tissue. This operation is regulated by specific physiological mechanisms—and *the environment influences these mechanisms*, in numerous ways. The manner in which the genotype expresses itself to form the phenotypic "you" is inextricably linked with the unique environment in which your development unfolded.

A simple example: memory In chapter 5, short-term memory was contrasted with long-term memory, and we explained that long-term memories are encoded in changes in the physical structure of nerve cells. We also said that synaptic connections multiply or die, depending on whether they are activated (used) or not. The structure of the connections within your brain therefore changes, in a simple and concrete way, as new long-term memories are formed.

This process involves the expression of genes. When one neuron activates another, it stimulates genes in the second neuron to manufacture particular proteins, which in turn leads to the growth of new synapses in that cell. It is difficult to imagine anything that is more environmentally determined than your autobiographical memory, yet its physical realization in the brain is mediated by the process of *gene* transcription. Many other psychological functions are molded in a similar way by the constant interaction of environmental and genetic mechanisms.

WHAT IS THE "ENVIRONMENT"?

When something acts upon a cell, from the cell's vantage point it does not matter whether the agent in question came from within the body or from outside it. The cell does not distinguish the source of a modification: from the cell's perspective, *everything* outside its small world is "the environment." If a metabolic process in the brain requires insulin, for example, it does not matter to the cell whether the pancreas produced the insulin or it was injected artificially into the blood stream. In either case, the necessary insulin influences the cell in the same way.

When *we* speak of the environment, we mean the outside world (everything outside our bodily selves). That is fair enough. But many people seem to think of external influences on development as *mental* and internal influences as *physical*. This can be confusing. The distinction between "mental" and "physical" is merely an artifact of one's observational perspective (see chapter 2). The action of cocaine stimulating the SEEKING system is something physical if you observe its effects through an fMRI scanner. However, it is something mental if you experience those effects from within, as an increase in your level of interest in objects in the world. *All* events are physical events when they are observed from the appropriate vantage-point, regardless of where they come from. In this chapter, we are considering environmental influences on mental development from the point of view of their impact on genetic mechanisms at the cellular level. The "environment" in question is therefore always *physically* mediated, although the *origin* of the environmental influences we shall discuss is always the *external world*.

CRITICAL PERIODS IN DEVELOPMENT

The intimacy of the link between genetic and environmental influences varies for different psychological functions at particular periods in the developmental process. The maturational sequence of the expression of genes in brain cells is associated with spurts in the production of synapses at different sites in the nervous system at different times. During these periods of rapid growth, many more connections are formed than will ultimately be used. The environment that the brain finds itself in at these critical times will determine which connections are used (are activated) and therefore which will or will not survive (see chapters 1 and 5). Those that are not activated sufficiently are "pruned" from the maturing structure. During these critical periods, therefore, maturing brain structures are particularly sensitive to environmental influences. There are a multitude of these kinds of critical periods during the first thirteen years of life—from infancy to, approximately, puberty.

SEX DIFFERENCES

The subject of sex differences is particularly appropriate for our purposes. There would appear to be nothing that is more obviously "genetically determined" than the difference between boys and girls. Males and females are patently different from the outset in their physical anatomy, and the differences are clearly grounded in their different chromosomes. These major anatomical differences are most obviously linked with sexual reproduction, and they therefore seem to predict certain necessary differences in instinctual behavior. In fact, it is widely believed that boys and girls differ in many aspects of their behavioral, emotional, and intellectual dispositions. To what extent are these *psychological* differences genetically determined?[1]

There are 23 pairs of chromosomes containing our genes. Males and females share 22 of them; they differ systematically in only one pair: females (normally) have what is called an "**XX**" pair on the 23rd chromosome, whereas males have an "**XY**" pair. The difference is therefore in the one "Y" chromosome. It is interesting that society makes so much of the difference between men and women, when all it literally boils down to (in genetic terms) is a difference in one chromosome out of 46. It is obvious that there is a great deal more that we share than what distinguishes us.

Testes and ovaries

The "default" plan of the human body, including the brain, is *female*. Unless a specific factor, which we shall identify in a moment, is brought to bear during the maturation of the fetus,

[1] A readable account of much of the material covered in the remainder of this chapter can be found in LeVay (1994).

everyone would end up with a female-type body. The fetal organs that produce sex cells are known as **gonads**. These organs are the same in both boys and girls, until a specific moment in the maturational process. At this point, the Y chromosome exerts its decisive influence. A short sequence of genes on the Y chromosome produces a substance called **testes-determining factor**. This factor acts on the gonads and influences the transcription function of the genes in their cells in such a way that an organ that would otherwise naturally have developed into an ovary becomes a testicle.

With the omission of a few minor details that do not affect the main issues, this is the essence of the male–female distinction. As with other biological mechanisms that are thoroughly understood, it is possible to manipulate this mechanism experimentally. Specifically, the testes-determining factor can be artificially introduced during the critical period in the maturation of an XX (female) fetus.[2] In spite of the female chromosomal (genotypic) "blueprint" of every cell in its body, the result is that the animal thereafter develops as an anatomical (phenotypic) male. Conversely, if the production of testes-determining factor in an XY (male) fetus is inhibited, the animal will continue to develop along female (default) lines. This is the first and most decisive step in the development of sexual differences. And clearly, even at this basic stage, it is possible for environmental factors to alter the process in a dramatic way.

Testosterone

During the second trimester of pregnancy, when the testicles start to develop, their cells secrete the hormone called **testosterone**. Everything else that distinguishes males from females

[2]This experiment was first carried out in 1991 (for details see LeVay, 1994, p. 20).

in these early processes seems to arise from the effects of testosterone. Testosterone acts on a range of organ systems in the body—all those that have *receptors* on their cells that "recognize" it. Receptors are like small keyholes, located on the surface of cells. If a molecule has the right shape (if it is the right key) to fit a particular receptor site, it will attach to the cell and produce the relevant effect on its workings. Testosterone, which flows throughout the bloodstream, is one such key. Wherever it bumps into cells with matching keyholes, it triggers a sequence of genetic events in those cells.

There is the same number of receptors for testosterone, in the same locations, in (genotypically) male and female bodies. However, males possess testes—due to the process just described—which means that their bodies produce testosterone in abundance. As a result, far more testosterone receptors are activated in the male body than in the female body. The activation of testosterone receptors has different effects on the cells in the different organ systems (i.e., they have different transcription effects), resulting in myriad anatomical changes: the formation of the genitalia; the secondary sexual characteristics (e.g., breasts, body hair, and vocal timbre); and the overall shape and size of the body. All of these changes, then, despite their apparently fundamental nature, are determined by delicate chemical processes that can readily be environmentally manipulated.

Testosterone and its vicissitudes The biochemistry of testosterone is complex. It does not act *directly* on cells but must be *converted* into other substances before it activates the relevant cells to stimulate sex-related changes. An enzyme called **5-alpha-reductase** is one chemical responsible for converting testosterone. This enzyme transforms testosterone into **dihydrotestosterone**, and *this* is the substance that triggers the process of bodily masculinization. The female body will therefore only become a male body if enough of this converted testosterone is present in it. If this substance is not sufficiently present, then the body will continue with its original plan of developing into a female. The obvious implication is that anything that reduces the amount of the enzyme 5-alpha-reductase will block the conversion of testosterone and will thus inhibit masculinization. This, once again, opens the way for environmental influences that can override the genotype.

Let us imagine a body with the XY (male) chromosomal pattern. In this case, testes have already been formed by the testes-determining factor. The testes will transmit testosterone through the blood supply to other tissues in the body. The tissues with the appropriate receptors will "recognize" it. If, however, there is not enough 5-alpha-reductase to convert this testosterone, then the tissues will not masculinize, and the body (despite the presence of testicles) will develop along female lines. This body will have the XY (male) chromosomal structure and possess a decisive male anatomical organ (testes), but it will nevertheless develop female external genitalia, female secondary sexual characteristics, and an overall "female" shape and size.

This is perhaps seen most famously in the case of Olympic athletes. When it became possible a few decades ago to test for XY chromosomes, the International Olympic Committee decided to introduce such tests to prevent (larger, stronger) male athletes gaining an unfair advantage by masquerading as women. The tests were stopped when cases presented themselves in which women, who were demonstrably female in terms of their anatomy, nevertheless had a "male" (XY) genotype. These individuals were genetically male, but anatomically female. Their genotype did not coincide with their phenotype, presumably because testosterone conversion had been altered during a critical maturational period.

There are various ways in which this can happen. In the 1950s and 1960s, a drug called progesten was given to pregnant women to prevent miscarriage. A side-effect of this drug was that it suppressed the process of testosterone conversion (LeVay, 1994, p. 27). A condition called congenital adrenal hyperplasia also blocks this process. The fact that a person's genetic inheritance can be so dramatically modified challenges the widely held belief that genes determine our fate in an unmodifiable and predetermined way.

Sex differences in the brain

Later in the second trimester, another critical maturational sequence takes place. These changes occur just after the sequence described previously, which modified the sexual body. This second wave of changes "sex-modifies" the brain itself. Once again testosterone needs to be converted, this time by an enzyme called **aromatase**, which transforms testosterone into **estrogen**. Estrogen is a hormone that is naturally produced by the ovaries, but the same chemical is responsible (during this critical period) for masculinizing the brain. Environmental disruption of the key enzyme can, as in the case of 5-alpha-reductase, derail the entire process. As a result, it is possible to live in a male sexual body that contains a "female" brain.

There are definite, albeit subtle differences between male and female brains. One clear difference is that the male brain is larger. The difference appears to be proportional to the size of the rest of the body. This implies that the *average* man has a larger brain than the *average* woman. This is true for the other organs too—hearts, stomachs, and livers. Larger brain size does not imply greater intelligence—if it did, then large men would be more intelligent than small men![3] There are, however, two

[3] Intelligence depends on the *pattern* of connections between cells, not their number (except in the extreme cases of certain medical conditions, which produce profound learning disability).

striking differences between male and female brains that are not related to brain size. There are many more minor differences—but these are the ones that have been most thoroughly investigated.

Hemispheric asymmetry The first region that has shown a reliable sex difference is the **corpus callosum** (see Figure 7.1). This fiber bundle connects the left and right cerebral hemispheres (see chapter 1). The corpus callosum is proportionally *larger in the female brain* than in the male. The masculinization of the brain therefore seems to involve suppression of growth of these fibers. As a result, the left and right hemispheres of the (average) female brain are more intimately connected than are their male counterparts.

The larger corpus callosum is thought to result in *less lateral specialization in females* than in males. There is, therefore, usually more equipotentiality between the hemispheres of females. In males, there is a greater division of labor between the hemispheres (they "put all their eggs in one basket"). The functional consequences of this are thought to be, firstly, that women commonly have *superior language abilities* (they speak better, earlier, and more), whereas men generally have *superior visuo-*

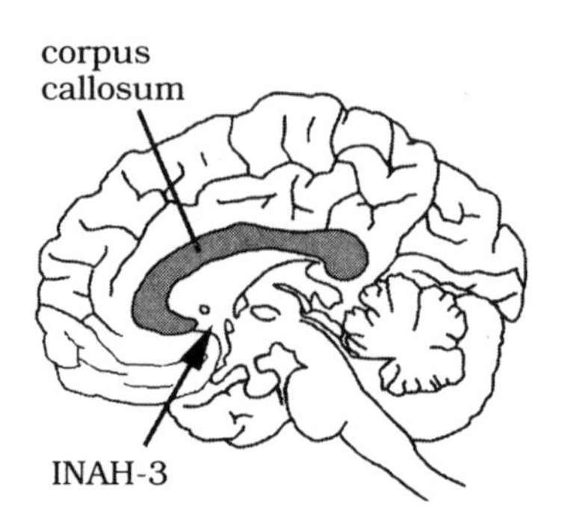

FIGURE 7.1
Corpus callosum and INAH-3

spatial abilities (such as the topographical skills required to remember a route or read a map). (For a review see Springer & Deutsch, 1998, pp. 139–156.) It is not entirely clear why the greater interaction between the hemispheres should lead to this pattern of differences. Nevertheless, the differences are well established and are the best-studied cognitive differences between males and females.

An important qualification in relation to these findings is that we are talking about the *average* performances of *large groups*. Average performances do not reliably predict the behavior of *individual members* of a group. Moreover, men and women do not differ dramatically in these respects; rather, they differ by *small* (but statistically significant) amounts.[4] Such cognitive differences are politically sensitive issues; it is interesting, therefore, that society makes so much of these minutiae. Clearly, the importance we attach to the differences is not rooted in *cognitive* factors.

While the differences in *cognitive* ability (and hemispheric anatomy) are vanishingly small, there are other differences between the sexes that are more substantial. These differences are reflected in our more primitive, subcortical anatomy.

The hypothalamus The second main neuroanatomical difference between the sexes is to be found in the medial preoptic area of the *hypothalamus* (see chapter 1). A set of nuclei known as the

[4] For the more statistically oriented: The size of the sex difference for such cognitive functions is "typically on the order of one-fourth of a standard deviation" (Springer & Deutsch, 1998, p. 156). By way of analogy, IQ tests are designed to have a mean of 100, and a standard deviation of 15. The differences we are speaking about here are such that if the average score of the "superior" group (say, women on a verbal task) was 106, then the "inferior" group would score an average of roughly 102 (i.e., four points—roughly one-fourth of the standard deviation of 15). Clearly, the effects are very small.

interstitial nuclei of the anterior hypothalamus (INAH) are located in this region. There are four of these nuclei, with visible sex differences in two of them. In one of the latter, the difference is small and has proved difficult for investigators to replicate (see LeVay, 1994, p. 76). In the case of the other, *INAH-3* (Figure 7.1), however, the difference is substantial—vastly greater than in the case of the corpus callosum. In this case, the male INAH-3 is very much larger than the female.

This sex difference is visible across the mammalian series, but the magnitude of the effect varies somewhat from species to species. In human males, INAH-3 is three times larger than in females. In the rat, it is five times larger. Although some of the findings discussed below come from animal experiments, the key findings of the greatest interest to us relate to humans.

The hypothalamus (of which INAH-3 is a small part) is the "headquarters" of the autonomic nervous system (see chapters 1 and 4). As such, the activities of the hypothalamic nuclei are intimately bound up with the hormonal economy. These nuclei constantly respond to, and alter, the levels of the various hormones coursing through our bodies. For this reason, sex differences in the hypothalamus have widespread ramifications elsewhere in the brain, and in the rest of the body. Most of the reproductive functions—which differ widely in the male and female (e.g., the menstrual cycle)—are regulated by hypothalamic nuclei. Possibly for the same reason, there are significant sex differences in brain chemistry. Specifically, the brain circuitry mediating sexual behavior (in the broadest sense) differs between the sexes. The sexual circuitry of women is mediated to a large extent by a peptide called **oxytocin**, whereas in men it is mediated largely by **vasopressin**. Other neurophysiological differences possibly related to this chemical distinction, and to the underlying hypothalamic differences, have been observed. These include the fact that the *anterior cingulate gyrus* is more active (in the resting state) in women than in men, and conversely that the *amygdala* is more active (in the resting state) in men than in women. In chapter 4 we learned that the anterior cingulate gyrus is a key component of the PANIC (separation-distress) emotional system. The greater activation of this region in females therefore seems likely to be linked with the expression of *nurturant* behaviors and concern for *social* connections. The amygdala, by contrast, is a key component of the RAGE emotional system. Its greater activation in males probably relates to the finding that boys are typically more *aggressive* and *active* than girls. These sex differences are not unique to humans. Male primates and rodents both display greater concern with power relations and dominance behaviors than do females. On the other side of the spectrum, one sees greater social concern and nurturant behavior in female primates and rodents than in males (see Panksepp, 1998).

"Mismatches" between body and brain

This, then, is what we mean by "masculinization" of the brain. The masculinized brain differs from the female brain in these respects. Remember, all of this is due solely to the conversion of circulating testosterone (by the enzyme aromatase) into estrogen, during a certain critical period of fetal maturation, after the sexual characteristics of the rest of the body have already been determined.

Recall also that this process is open to environmental manipulation. When the action of aromatase is suppressed in male rats, for example, their brains fail to masculinize. As a result, instead of displaying the "male-typical" behaviors mentioned above, they display "female-typical" behaviors—including sexual behavior such as exposing their genital area for the purposes of penetration, instead of mounting the female. This follows, literally, from having a female brain in a male body.

These effects are not restricted to rats. In a family in the Dominican Republic, XY males with a rare disorder develop female bodies despite having male brains. The situation alters at puberty, when their bodies belatedly masculinize. Since they appeared (externally) until that moment to be female, their unsuspecting families had raised them as girls, until the surprise of puberty. This permits some interesting observations as to the extent to which social factors determine gender identity. For the most part, these misidentified "girls" readily switch to being boys and live happily as men after puberty (see Rogers, 1999, pp. 33 ff.).

A second example demonstrates the point more clearly. A small number of normal XY males have been accidentally castrated—often due to surgical errors in circumcision. Doctors typically advise that such a child be given female-like genitalia, that "his" hormonal balance be altered through medication to that of a female, and that he be raised as a girl. Everything may then seem to be fine until puberty, notwithstanding "tomboyish" behavior. However, at puberty the child typically experiences a crisis of gender identity. In one well-publicized case, a boy had his surgery reversed on reaching adulthood. His external genitalia now look male but are not fully functional. He nevertheless lives happily as a man (see Rogers, 1999).

The common thread that runs through all of this is that it is the sexual identity of the *brain*, whether it is feminine or masculine, that appears to determine whether people "feel" female or male and behave in a "typical" feminine or masculine way.

Environmental influences on sexual orientation: The example of prenatal stress

Readers might understandably complain that the environmental influences we have discussed up to now are far removed from the sort of "life-event" factors that typically concern psychotherapists. It is therefore important to remember what we said in this respect at the outset: from the viewpoint of neurophysiology, all "life events" are ultimately mediated (registered and translated) by bodily events. *Stress* provides a good example.

There is abundant evidence (in the animal literature) that sexual orientation can be determined by environmental stressors during critical maturational periods. In rats, the critical period in question occurs just before birth. An increase in *maternal* stress during this period causes a premature surge of testosterone *in utero*, with the result that the brains in the (already masculinized bodies) of their XY male fetuses *fail to masculinize*. The stress is induced in the pregnant rats by overcrowding the cages in which they live or by producing frequent (weak but unpredictable) electric shocks to the base of the cages. When the male pups are born, only 20% of them show active sexual behavior of any kind (80% of rats are normally sexually active), and a full 60% of those that are sexually active display "female-typical" sexual behaviors (which in rats involves arching the back and exposing the genitals). This group also includes some rats who are (for want of a better term) "bisexual"—displaying both male-typical and female-typical behaviors.

Given what we know about the preservation of these basic mechanisms in the mammalian series, similar effects may well occur in humans. The critical period for brain masculinization in humans is earlier than in rats—it occurs in the second trimester of pregnancy. It is difficult to be precise about the effects of prenatal stress during this period in humans, because we cannot bring the same degree of experimental control to bear on our studies. One study that attempted to address the issue compared the number of homosexual men as against heterosexual men born in Germany (a) before the Second World War, (b) during and immediately after the war, and (c) well after the war (Dorner et al., 1980). The hypothesis that group (b) would contain a greater proportion of homosexuals (having been exposed to greater prenatal stress) was confirmed by the results of this study. However, the effect may have been caused by factors other than prenatal stress.[5]

In any event, it seems unlikely that sexual orientation in humans is so simply determined. Moreover, masculinization of the brain presumably affects *gender identity* rather than *sexual orientation,* which may have a more complex determination.

Perhaps the most famous neuroscientific finding regarding human homosexuality is that of LeVay (1991; 1994, pp. 120–122), who compared the size of the interstitial nuclei of the hypothalamus in homosexual and heterosexual men. He focused specifically on INAH-3 (see above) and found that, in homosexual men (who had come to autopsy as a result of AIDS), INAH-3 was three times smaller than in heterosexual males. As mentioned earlier, this nucleus is typically three times smaller in the female than in the male. There were no differences in the other INAH structures. This finding provides further evidence that INAH-3 is related to sexual difference—perhaps particularly in relation to the *aim* of the sexual drive. It is not clear how environmental factors, including the effects of stress in the second trimester, might influence the size of INAH-3. A number of hereditary and environmental factors may have a significant influence on the size of INAH-3, and INAH-3 might well not be the only neuro-anatomical predictor of male sexual orientation.

Hereditary influences on sexual orientation: A "gay gene"?

The standard way of ascertaining the extent of hereditary influence is to look at the prevalence of a certain behavior in monozygotic twins who are reared apart from one another.

[5] LeVay (1994, p. 125) points out: (1) it is difficult to reliably establish the incidence of homosexuality in different age groups; (2) there may have been influences other than stress (e.g.fathers would typically have been absent from the family home during the war). (See also Bailey, Willerman, & Parks, 1991.)

Monozygotic twins have an identical genetic make-up and are thus commonly referred to as *identical twins*. By measuring the level of agreement (or "concordance rate") between the behavior of these twin-pairs, it is possible to crudely estimate the extent of the contribution of heredity.

In male homosexuals, this concordance rate is roughly 50%, and in female homosexuals it is roughly 30% (the incidence of homosexuality in the general population is roughly 10% or less: LeVay, 1994). Thus, there appears to be a significant hereditary contribution to homosexuality, and this is likely to be mediated by genetic mechanisms resulting in anatomical differences such as that in INAH-3. In this context, it is relevant that a specific gene sequence that may be associated with male homosexuality has been identified (Hamer, Hu, & Magnuson, 1993). This sequence has been labeled **Xq28**. (The fact that the sequence is on the X rather than the Y chromosome is consistent with the truism that the pattern of inheritance runs down the female line.) When this finding was published, there was a great deal of media coverage about the discovery of a "gay gene." Since then, there has been considerable uncertainty about the reliability of Hamer's findings. But, assuming that it is eventually confirmed, it is important to remember that it is a *sequence* of genes, and this sequence of genes can only be *one of the factors* that determine sexual orientation. Even in the subset of homosexual males in whom the gene sequence is present, it is likely to interact with environmental events in many complex ways.

To illustrate the point: the conclusion that 50% of the variance in male homosexuality (30% in female homosexuality) is determined by heredity can also be reversed. What is it that determines the other 50% (or 70%) of the variance? Why are genetically identical twins not *always* identical in their sexual orientation? Many complications arise with questions of this sort, not least of them being the fact that twins share the same *intrauterine environment*, which appears to be a significant factor when it comes to sexual orientation. Even more difficult to disentangle are the environmental influences that genes *bring upon themselves*, as it were.

The multiplier effect

We have said already that the genetic differences between males and females are minuscule. The anatomical and physiological differences arising from those genetic differences are also extremely small in comparison with the commonalities. The psychological *effects* of these small differences, however, are *multiplied* during development. The simple effects of adults' expectations (which themselves have a complex causation) are widely recognized. For example, caregivers talk more to babies dressed in pink romper suits than to those in blue, and they interact more physically when the same babies are dressed in blue romper suits (see Rogers, 1999). Less widely known is the "multiplier effect." Children who are, for example, innately more active and aggressive, as opposed to nurturing and sociable, will literally *create* different environments for themselves. Not only do caregivers respond differently to male and female children, male and female children also *elicit* different types of responses from caregivers (and the rest of the world) by virtue of their male-typical and female-typical behaviors. These different responses, in turn, stimulate further differences in the original behaviors, and so on. Thus, although the innate differences might be tiny to begin with, they rapidly self-propagate and expand. In this way the genetic differences cause environmental differences, which then become impossible to disentangle from one another (when calculating "concordance rates," for example).

Here are some final—striking—examples of the complex interaction between genetic and environmental factors in the development of sexual differences.

Maternal influences on sexuality

Mother rats have a peculiar habit of licking the anogenital area of their male pups (for details of the following observations see Rogers, 1999). This appears to encourage male-typical behaviors in the pups. The mothers lick their male pup's anuses and genitals because testosterone, interacting with other chemicals, generates a smell in the anogenital area that female rats find attractive. This mechanism is demonstrated when the smell is suppressed and the licking then ceases. Similarly, artificially introducing the chemical into the anogenital area of female pups causes the mothers to lick them too. This triggers a cascade of male-typical changes in the female pups, including increased size of INAH-3 and mounting-type sexual behaviors. The converse applies to the male pups: after suppression of the licking behavior, they develop female-typical brain morphology and female-typical sexual behaviors. It is most important to note that these changes are caused by the *licking*, not by the underlying hormone that (normally) stimulates it. Here, then, is an example of testosterone-induced neuropsychological changes being mediated not directly, but by the environmental responses that a hormone elicits.

A final observation refers back to one of the studies mentioned previously, in which female-typical sexual behavior was induced in XY male rats by stressing their mothers during late pregnancy. These findings were followed up: Half the "homosexual" rats were reared by sexually active adult females (not necessarily their biological mothers, but adult females who were sexually experienced). The other half were reared by sexually inexperienced adult females. The result was an enormous difference in the subsequent sexual development of the two groups. The incidence of "male-typical" sexual behaviors increased to 50% in the first group, whereas only 2% of the rats in the second group ever developed male-typical sexuality.

Even though *human* mothers are not inclined to lick the anogenital area of their infants, it is quite likely that they do also interact with their male and female babies in different ways. As suggested previously, studies indicate that mothers show some types of physical contact with the bodies of their male babies significantly more frequently than with their female ones. This may well promote differential morphological changes, perhaps analogous to those seen in rats, in the babies' brains.

CLOSING REMARKS

This chapter has attempted to demonstrate that environmental and genetic influences are *absolutely inextricable*. The genotype (the design according to which you are built) is open to a wide range of manipulations, as it expresses itself in a particular environmental context, which in turn shapes the phenotype ("you" yourself). Most people are of the opinion that sex and gender—and all that they imply—are predetermined from the moment of conception by our genetic make-up. It is our hope that this chapter has convinced readers that sexual development is not that simple. And if we have succeeded in demonstrating this in the case of sex differences, we hope that readers will be prepared to extrapolate the principles to other constituents of the inner world of the mind, in which the impact of the environment cannot be any less decisive.

CHAPTER 8

WORDS AND THINGS: THE LEFT AND RIGHT CEREBRAL HEMISPHERES

The great cerebral hemispheres of the forebrain have been mentioned frequently in previous chapters, but mainly to contrast them with the deeper brain structures that have been our main focus. This chapter deals exclusively with the higher forebrain and, more specifically, with functional differences between the left and right cerebral hemispheres. Over the years, the functional asymmetry of the cerebral hemispheres (unlike just about everything else about the brain) has attracted some interest from psychoanalysts. In this chapter, alongside our review of the basic facts of functional cerebral asymmetry, we comment on what these psychoanalysts have made of these facts. This will pave the way for the final two chapters of our book, in which we intend to deal with psychoanalytic matters in more depth. We begin, then, with a review of some of the basic facts concerning the functional differences between the hemispheres.

HISTORICAL ORIGINS

Interest in the asymmetrical contribution that the two hemispheres make to our mental life can be dated back to Broca's celebrated case report of 1861, which we have mentioned more than once already. Readers will recall that Broca's patient "Tan-Tan"—who lost the power of speech after a stroke—suffered damage to the left-hand side of his brain, mainly in the inferior, posterior part of the frontal lobe (now known as **Broca's area**). Four years later, Broca described a larger group of cases with similar disorders, all of whom had lesions in roughly the same area. In fact, it was only *then* that Broca realized that it mattered which *side* of the brain was damaged. The idea thus arose that language was bound up with the functions of the *left* cerebral hemisphere. Broca also suggested that the leftward lateralization of language might be related to the fact that most humans are right-handed (and therefore that the *right* hemisphere might be dominant for language in left-handers). The relationship between handedness and hemispheric dominance for language turned out to be slightly more complex (for a review see Springer & Deutsch, 1998). For the purposes of this chapter, we will consider only the simple case of "typical" hemispheric asymmetry (found in almost all right-handers).

REGIONS THAT SHOW HEMISPHERIC ASYMMETRY

The fact that the two hemispheres are almost identical anatomically is no less striking today than it was in Broca's time. Even though a number of minor anatomical differences have been discovered over the years (see Springer & Deutsch, 1998, ch. 3), these are very subtle, and they amount to minor distortions in an otherwise mirror-image pattern. As regards mental functioning, in contrast, the two hemispheres are radically different.

This is not true for all parts of the hemispheres. The functional properties of the "primary" cortical zones (see Figure 8.1)—where visual, auditory, or somatosensory information are "projected" onto the cortex (see chapter 1)—are completely symmetrical. The right visual field simply projects to the left visual

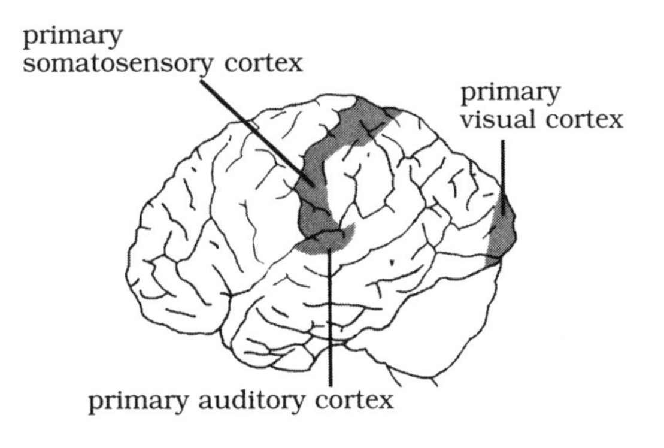

FIGURE 8.1
Projection cortex

cortex, and vice versa; sensory information from the right half of the body projects to the left somatosensory cortex; and so forth. The same applies to the "action" end of the system: primary *motor* cortex is symmetrically organized—the primary motor cortex of the left frontal lobe controls the movements of the right half of the body, and vice versa.

Asymmetry only arises at the level of the secondary and tertiary cortical zones, otherwise known as "association" cortex (see chapter 1). This asymmetry is apparent in two large areas on the surface of the hemispheres (see Figure 8.2). The first is broadly in the area of the occipito-temporo-parietal junction, toward the back of the brain. The second is almost the entire

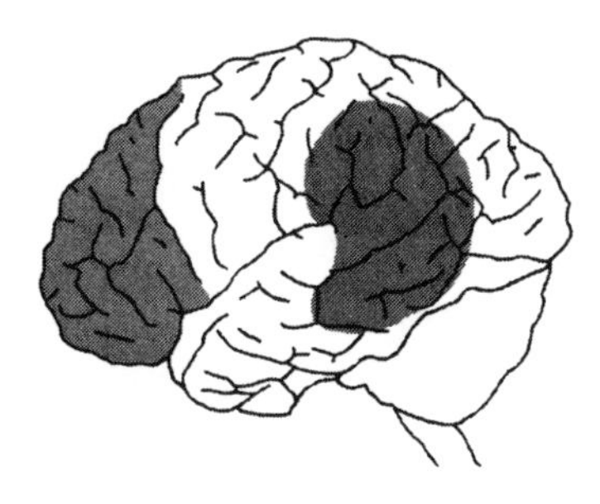

FIGURE 8.2
Association cortex

prefrontal area. When we speak of the asymmetrical functions of the hemispheres, it is primarily to these two parts of the hemispheres that we are referring.

THE ASYMMETRY OF MENTAL FUNCTIONS

Broca's initial observation that the "seat" of language was leftwardly lateralized was a momentous discovery. It gradually became apparent in the years following Broca's discovery that several different types of language disorder arose from lesions to different parts of the left hemisphere. For example, Carl Wernicke observed in 1874 that only speech *production* was affected by damage to Broca's area; speech *comprehension*, on the other hand, was affected by damage to the superior, posterior part of the left temporal lobe (which later became known as **Wernicke's area**). Disorders of other cognitive functions, such as reading, writing, and arithmetic, were also documented and localized to other specific regions of the left cerebral hemisphere. The fact that some of these left-lateralized functions were not obviously language-based[1]—and, no doubt, also the fact that most people are right-handed—led to the conclusion that the left hemisphere was somehow *dominant* over the right.

This view gradually receded as we realized that both hemispheres were "dominant" (or, more appropriately, "specialized") for *different functions*. For example, patients with right-hemisphere damage seemed to have particular difficulty in copying drawings, and with constructional tasks in general. They also seemed to be especially disabled with respect to topographic orientation. By the middle of the twentieth century, it was generally accepted that, if *language* was the prime function of the left hemisphere, then the right was specialized for *spatial* cognition.

[1] For example, a system critically involved in *skilled movement* was identified in the left inferior parietal region.

But some findings did not seem to fit this neat dichotomy. We have already mentioned that skilled movement is bound up with the functions of the left hemisphere. Skilled movement does not sit easily under the heading of "language functions." Likewise, it became clear that the right hemisphere was also specialized for some functions that could not be described as "spatial." For example, patients with right-hemisphere damage have special difficulty recognizing and producing the "prosodic" (intonational) aspects of speech. Investigators now began to seek the *underlying common factor* that could account for the full range of these clinical findings.

Grand theories of lateral asymmetry

Various generalizations have been put forward (for a review see Springer & Deutsch, 1998, pp. 292–301). For example, it has been suggested that the left hemisphere is specialized not for language *per se*, but for a more fundamental function upon which language is dependent, such as *analytical* or *sequential* information processing. The right hemisphere, by analogy, is argued to be specialized not for spatial cognition as such but, rather, for *holistic* and *simultaneous* processing. Table 8.1 provides a representative list of such functional dichotomies.

TABLE 8.1. Left–right dichotomies

"Dominant" (left) hemisphere	*"Minor" (right) hemisphere*
Verbal	Nonverbal or perceptual
Symbolic	Visuospatial
Verbal	Visuospatial
Analytic or logical	Perceptual or synthetic
Propositional	Appositional
Analytic	Holistic
Propositioning	Visual imagery

During the 1960s and 1970s, these ideas began to enter popular culture (see Springer & Deutsch, 1998, pp. 289–301). It was claimed that the left hemisphere embodied rational and logical thinking, while the right hemisphere was more intuitive and creative. Soon we were being told that many people use only "one half" of their brains, typically the left half (which was presented as being "logical," "rule-governed," and "uncreative"). Before long, the scientific and technological achievements of industrialized Western nations (contrasted with the elusive, mystical thinking of Eastern religions) were being attributed to functional hemispheric asymmetries! There is little empirical support for such ideas.[2] Nobody who has worked clinically with the tragic consequences that *really* arise when a patient loses the functional contribution of half of his or her brain (or even a small part of it)—as occurs with strokes, for example—will take any of these propositions seriously.

All of the attempts to dichotomize the basic mental functions of the left and right hemispheres have proved futile, and it is likely that there is no single fundamental factor that distinguishes the functions of the two hemispheres. The fact that they are *anatomically* dichotomous does not imply that they must be *functionally* divided in an equally simple way. In reality, the functional distinction between the hemispheres is multifactorial. The hemispheres differ in many respects, some related to each other and some unrelated. Moreover, almost all mental functions (as we know and classify them in psychology) incorporate functional contributions from *both* cerebral hemispheres. In fact, the hemispheres do not contain "mental functions" as such. Rather, different parts of the different hemispheres are recruited into the *complex functional systems* that mediate all our mental faculties (see chapter 2). It is a rather complicated state of affairs, one that does not lend itself easily to simple arguments.

[2] Indeed, the section of Springer and Deutsch (1998) just referred to earlier is entitled "Hypotheses and Speculation: Beyond the Data"!

PSYCHIATRY, PSYCHOANALYSIS, AND HEMISPHERIC ASYMMETRY

Sadly, simple arguments concerning hemispheric asymmetry found a receptive audience in the psychoanalytic community. One claim that has been made repeatedly in the psychoanalytic literature (for a review see Kaplan-Solms & Solms, 2000) can be traced back to a paper published by David Galin in 1974. Galin claimed that the left hemisphere—the verbal, analytical, logical side—is the seat of Freud's system *consciousness*, with its "secondary-process" mode of thinking. By contrast, the right hemisphere—the concrete, holistic, intuitive side—is the seat of the *unconscious* and therefore of primary-process thinking.

The psychoanalysts who have repeated and propagated this assertion have almost invariably used arguments by analogy to make their case, basing these on generalizations about the supposed underlying hemispheric functions that were tabulated above. The typical argument, first propounded by Galin (1974), is that since the system *Ucs* and the right hemisphere both think nonverbally and illogically, they must be one and the same, and *mutatis mutandis* with respect to the system *Cs* and the left hemisphere.

Split-brain studies

It is not difficult to see where Galin got these ideas from. So-called "split-brain" studies (mentioned briefly in chapter 3) were extremely influential in the 1960s and 1970s. At that time, a surgical procedure had been introduced for the treatment of intractable epilepsy: the corpus callosum (the fiber band that connects the two hemispheres; see chapters 1 and 7) was cut to isolate the seizures and prevent them from spreading to the healthy hemisphere (see chapter 6).

This operation (known as "**commisurotomy**") was performed for purely clinical reasons, but it provided a unique *scientific* opportunity for neuropsychologists to study the functions of the two hemispheres independently. However to investigate the problem systematically, some methodological problems had to be overcome. The main problem arises from eye movements: if the eyes can move freely when looking at objects placed in front of them, both hemispheres receive information about the objects. This can be controlled by asking the patient to fixate on a spot and then briefly presenting the desired stimulus to one visual field while obscuring the other, thereby restricting the information to one hemisphere. Another problem is that the patients who underwent commisurotomy did so because their brains had preexisting abnormalities. The performance of "split-brain" patients therefore presumably does not reflect the functioning of neurologically normal individuals. This can be controlled by combining the results of split-brain studies with those derived from other methods and accepting only the converging evidence. Every scientific method has its limitations.

In split-brain studies, then, the isolated left hemisphere would be presented with (for example) images of printed words, such as "PEN" or "GLOVE," and the patients would be able to read these words. However, when the same images were presented to the isolated right hemisphere, the patients were not able to decipher them. The patients would recognize objects that had been named visually to the isolated right hemisphere (e.g., they could select the appropriate concrete object from a multiple-choice array presented to the left hand), but they would not be able to *name* verbally the objects thus recognized.

Using this paradigm, emotionally arousing images were presented to the isolated right hemisphere, eliciting a response that the left hemisphere was at a loss to explain. We described an example of this type of experiment in chapter 3: Pornographic pictures were presented by a male examiner to the right hemisphere of a female patient. She blushed and giggled accordingly, but she was unable to identify the source of her embarrassment. Interestingly, in such situations, split-brain patients frequently fabricate (or *confabulate*) explanations for their behavior. These explanations are highly reminiscent of what psychoanalysts call "rationalization"—that is, the patients generate plausible post-hoc motivations to justify behaviors that (as the examiner knows) were really motivated by something quite different.[3] Although the right hemisphere is "unconsciously" aware of the real motive, the "conscious" left hemisphere is oblivious of it.

Here, the term "*conscious*" is being used as if it were synonymous with *reflexively* conscious (see chapter 3), hence the quotation marks. Being aware of something (core consciousness) and being able to reflexively recall it and articulate it (extended consciousness) are not the same thing. We shall return to this point later.

The right hemisphere and the unconscious

When the psychoanalytic authors mentioned above claimed that the left and right hemispheres embody Freud's systems *Cs* and *Ucs*, respectively, they did not concern themselves with such complications. They interpreted the dissociated behavior of split-brain patients as a literal disconnection between the systems *Cs* and *Ucs*—that is, they interpreted it as artificially induced *repression*. By reverse logic, they believed (indeed, their own assumptions required it) that *normal* repression actually involves a functional disconnection between the hemispheres. The

[3] Similarly, in cases of posthypnotic suggestion, subjects are asked to perform a task while in a trance. When they are brought out of the trance, they carry out the requested behavior, concocting some rational-sounding (but false) post-hoc motivation to explain it.

corpus callosum becomes, in this way, the organ of repression. Before we can assess this startling conclusion in the light of appropriate empirical evidence, we need to introduce readers to some further neuropsychological complexities concerning the asymmetrical functions of the hemispheres.

MORE ABOUT THE NEUROANATOMY OF LANGUAGE

On the basis of lesion studies, and more recently through functional-imaging techniques, we have a thorough understanding of the anatomical organization of what Freud (1891b) called the "speech apparatus."[4] Speech information arrives in *primary auditory cortex* (*a* in Figure 8.3). This is the region that registers sound. Directly adjacent to this region is the *auditory association cortex* (*b* in Figure 8.3). This region chunks sound into recognizable units. Only some sounds are recognizable speech-sounds; these are called **phonemes**. All language is structured out of phonemes, but different languages recognize different phonemes. For example, Xhoi speakers (a Southern African language) use a variety of *click* sounds to convey meaning verbally, whereas clicks are never used in spoken English. Similarly, English speakers distinguish between "l" and "r" sounds, but this distinction is not recognized in Japanese. Damage to this part of the brain in the left hemisphere undermines the phoneme-recognition function, resulting in a speech-comprehension disorder mentioned previously: *Wernicke's aphasia*. These patients are able to produce language but cannot understand what they hear, and because their own language output is

[4] For detailed reviews of the modern neuropsychological concepts, see Bradshaw and Mattingley (1995), Feinberg and Farah (1997), Heilman and Valenstein (1985), Kolb and Wishaw (1990), Luria (1973), McCarthy and Warrington (1990), or Walsh (1985).

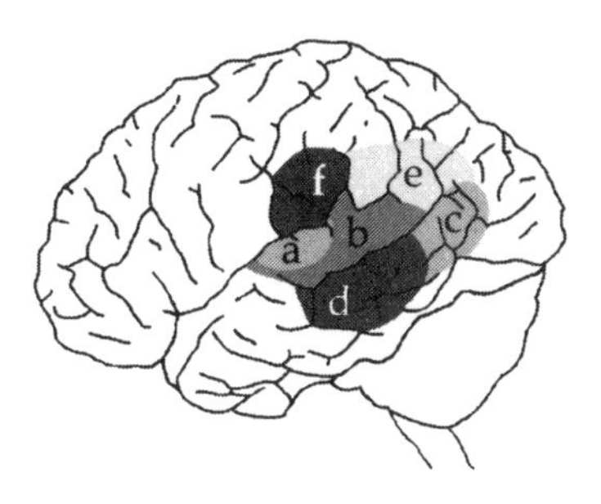

FIGURE 8.3
Sensory language areas

not appropriately modulated by auditory feedback, their speech is also abnormal.

To decode incoming audioverbal information, not only *phonemes* but also whole *words* and *sentences* have to be stabilized in short-term memory.[5] Words are clusters of phonemes that, partly by association with visual information (encoded in *occipito-temporal* structures; *c* in Figure 8.3), evoke concrete images in the listener's mind—thereby rendering the sounds meaningful. This is the **lexical** (as opposed to the **phonological**) level of analysis. Lesions affecting this system cause **anomia**—or, in more severe cases, **amnestic aphasia**—an extreme form of the "tip-of-the-tongue" phenomenon, in which one knows the meaning of the word one intends to use but cannot retrieve its phonology. The same thing can happen in reverse: patients with damage in this general area register the phonological structure of words, but they cannot retrieve the associated meaning (this is called *alienation of word meaning*).

The further stabilization of whole *sentences* recruits audioverbal STM mechanisms in the *midtemporal region* (*d* in Figure 8.3). It is often necessary to store a good many words in short-

[5]The STM systems are organized along modality-specific lines, as discussed in chapter 5.

term memory before the meaning of a spoken sentence becomes clear (consider the example of German, where the verb frequently comes at the end of the sentence). Lesions in the midtemporal region (and environs) produce a syndrome known as *acoustico-mnestic aphasia* or (more commonly) **conduction aphasia**. These patients register what is said to them at the phonological and lexical levels, but they cannot stabilize whole sentences reliably enough either to decode the meaning or to *repeat* the sequence accurately.

Moving upward toward the *parietal cortex* (*e* in Figure 8.3), and therefore toward the more abstract levels of the visual system (see chapter 6), there is a mechanism that deals with the analysis of the **syntactical** structure of speech. This is, in effect, a spatial (or, more properly, a quasi-spatial) mechanism. (The fact that *spatial* aspects of language are analyzed in a region of the *left* hemisphere demonstrates once more the foolhardiness of assigning complete mental functions to the different hemispheres.) Lesions here make it difficult for patients to decode the meaning carried by the quasi-spatial structure of word sequences. This is best conveyed by two examples: these patients have difficulty distinguishing between "the cat chased the rat" and "the rat chased the cat," or recognizing the different meaning of the phrases "my father's brother" and "my brother's father." The meaning of these phrases is wholly dependent on the *relative locations* of key words. Disorders of this function are known variously as **transcortical sensory aphasia**, or (less commonly) *receptive agrammatism*, or (confusingly) *semantic aphasia*.

All of these "sensory" aspects of language are mediated by components of the functional unit for receiving, analyzing, and storing information (see chapter 1). The "motor" aspects of language are mediated by the more anterior parts of the left hemisphere—in the frontal unit for the programming, regulation, and verification of action (see chapter 1).

The motor aspects of language

In reality, the "sensory" and "motor" aspects of language are no less inextricable than the functions of the left and right cerebral hemispheres. For example, even the elementary business of articulating words involves constant feedback from the somatosensory system regarding the changing positions of the lips, tongue, and pallet. This function is performed by a region of the *parietal* lobe (which is a *sensory* region; *f* in Figure 8.3), and lesions here contribute to the syndromes of *conduction aphasia*, described above, and *Broca's aphasia*, described below. (Luria identified a "sensory-based" motor disorder, which would account for some cases classified as Broca's aphasia, by the name *afferent motor aphasia*.)

Our description of the more strictly "motor" aspects of speech begins with the *intention* to speak. This complex function obviously cannot be localized in any narrow sense, but it is traditionally attributed to the prefrontal cortex (*a* in Figure 8.4). Disorders of speech initiative are commonly seen with damage to a part of the frontal lobe known as the *supplementary motor area* (*b* in Figure 8.4). The resultant disorder is known as **transcortical motor aphasia** (or *dynamic aphasia*, according to Luria's nomenclature).

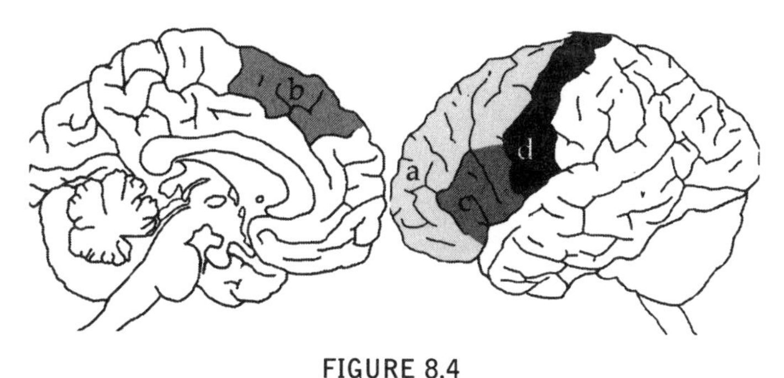

FIGURE 8.4
Motor language areas

The intention to speak results in the formulation of an actual utterance—which we can describe as a verbal motor program, or an articulatory sequence. The key region here lies farther back in the frontal lobes, in the *premotor* region, which includes Broca's area (*c* in Figure 8.4). Patients who suffer lesions here retain the ability to initiate speech, but they are unable to transform the intention into a concrete motor program. This is **Broca's aphasia** (of which "Tan-Tan" suffered an extreme form). (Luria identified this as *efferent motor aphasia.*) These patients can utter individual words (often only a few stereotyped words are available to them, mainly nouns), but they cannot string them together in grammatical sequences.

The complex sequence of planned articulations is then transformed into actual muscular activity in the *primary motor cortex*, in the most posterior parts of the frontal lobe. The specific region concerned is the most inferior part of the motor strip (*d* in Figure 8.4), which is dedicated to moving the facial and oral musculature. Lesions here produce unilateral facial paralysis—not an aphasia (remember: only association cortex is functionally asymmetrical).

The reader should now have a reasonably comprehensive picture of how speech is produced and analyzed in the brain. But communication is only one function of language. In this regard, Luria distinguished the *communicative* function of language from its other functions, such as its function as an "*intellectual tool*" and its role in the *regulation* of action.

BEYOND COMMUNICATION

Freud argued that attaching words to thoughts makes it possible to bring those thoughts to consciousness. In fact, in a paper called "The Unconscious" (Freud, 1915e), he explained that this was the rationale for his "talking cure." Repression involves the withdrawal of verbal associations from motivational programs, and the talking cure reconnects them. Repressed wishes are therefore literally unthinkable because they are unspeakable.[6] Freud believed that only something that has been *perceived* can become conscious. This is because consciousness is a perceptual function (see chapters 2 and 3). Unconscious mental processes far removed from the perceptual periphery—such as the deep motivational processes mediated by the "need detectors" of the hypothalamus and related structures (described in chapter 4)—cannot be rendered conscious *until they are associated with something perceivable*. Since the memory traces of words—what Freud called "word presentations" [*Wortvorstellungen*]—are derived from conscious auditory and kinesthetic sensations, they possess the requisite perceptual properties. Unconscious thoughts can therefore be made conscious by representing them in words.

Freud acknowledged that "word presentations" were not the *only* route from the unconscious to consciousness. Unconscious thoughts (what Freud called "thing presentations" [*Dingvorstellungen* or *Sachvorstellungen*]) can also become conscious in the form of concrete images ("object presentations" [*Objecktvorstellungen*]), such as the visual images that occur in dreams. They can also be *felt* consciously as emotions (see chapters 3 and 4). However, according to Freud (1923b, 1940a [1938]), language provides the most efficient and flexible means of representing our thoughts to ourselves.

What happens to a patient who loses this important function of language?

---

[6] Freud himself later realized that the process was, in fact, more complicated (Freud 1923b, 1940a [1938]). We have discussed some of the complexities of repression already, in chapter 5. See also chapter 9, where we discuss the "talking cure" in detail.

A patient who loses her thoughts

Karen Kaplan-Solms treated a patient, Mrs. K (see Kaplan-Solms & Solms, 2000, pp. 90–115), who sustained a hemorrhage in the midtemporal area of the left hemisphere (Figure 8.5). Initially, when Mrs. K awoke in hospital, she suffered from Wernicke's aphasia. She felt as though everyone was speaking a strange, unfamiliar language that she could not understand. She momentarily feared that she might be in Heaven, especially as she began to recall what had happened to her. However, she rapidly made better sense of her environment. Although she could not understand what anyone *said* to her, it was evident from the *appearance and behavior* of the people around her (nurses, doctors, other patients) that she was in a hospital. Mrs. K's phonemic hearing soon recovered, and she began to comprehend what

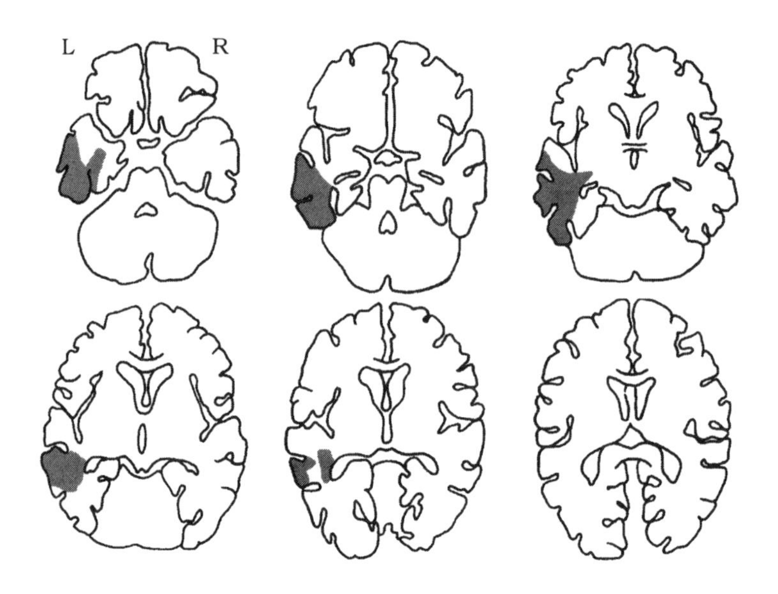

FIGURE 8.5

Lesion in patient losing thoughts

This figure is based on CT scans and shows sequential horizontal slices through the lower part of the brain (top left slice in figure) toward the vertex (bottom right slice in figure)

was said to her, so long as people spoke in short sentences. She was now suffering from a residual disorder of audioverbal short-term memory, causing acoustico-mnestic aphasia. As a result, she was unable to hold in mind for more than a brief moment anything that people said to her.

This was associated with a curious subjective state. Mrs. K kept "losing" her thoughts. A thought would occur to her, but before she was able to do anything with it, it was *gone.* Just as she was unable to hold on to what other people said to her, so too she was unable to retain what she "said" to *herself.* It was as if her consciousness had become a sieve. The same thing happened when she tried to converse with other people. She would formulate the words that she wanted to say, but before she could utter them they had vanished, leaving her speechless and confused.

The severity of this condition fluctuated. Occasionally, Mrs. K noticed that her whole mind had gone "blank"—all her thoughts were lost—not just those related to things she had heard or wanted to say. This state of mind, in which she could not think consciously of anything, was understandably frightening and embarrassing. She responded by retiring to bed and waiting for her thoughts to "come back," which they typically did after some time. When Mrs. K was at home during weekends, she would frequently withdraw from social interactions and sit quietly in a private place, such as her bedroom, waiting for her "mind to come back," as she put it.

In cognitive terms, it is understandable that her thoughts would disappear in this way. This patient sustained damage to the midtemporal region of the left hemisphere—which normally holds strings of words in short-term memory. Damage to this system affects not only the ability to hold in mind the words that one hears, but also the words that one generates in one's own consciousness. This is because the same audioverbal "buffer" is used for words that are generated internally as for words that are externally perceived.[7] Since the patient's audioverbal system could not contain her internally generated thoughts in short-term memory, these thoughts would disappear. This seems to confirm the proposal of Freud (and of many others) that we communicate our thoughts to our (conscious) selves by clothing them in words.

According to Galin's hypothesis (1974), however, loss of the capacity to think in words should result in nothing short of a psychotic illness. If the ability to think in words provides the very fabric of the system *Cs* (the secondary process in general), then destruction of that function should be tantamount to a destruction of all ego processes, leaving such patients at the mercy of their primary processes (and the instinctual drives that dominate the system *Ucs*).

Mrs. K's psychiatric status Did Mrs. K develop psychotic delusions? No. There is abundant evidence that her ego functions were fundamentally intact: despite her difficulties, her behavior continued to be governed by secondary-process thinking and the reality principle. For example, she tested her (momentary) delusional belief that she was in Heaven against the evidence of her external perception,[8] and this mental work resulted in the subordination of her phantasies to realistic perceptions. Similarly, when she "lost" her thoughts, she was rational enough to retire to her bedroom and wait for her "mind" to return. This seems a perfectly sensible solution to the problem. It is obvious that this patient had not *really* "lost her mind"; all she had lost was the capacity to *represent* (or retain) her thoughts in extended consciousness. Her mind (her ego)[9] continued to exist,

[7] We observed an analogous situation regarding the *visual* STM buffer in dreams (see chapter 6).

[8] Note that in such cases *core* consciousness remains intact (cf. chapter 3).

[9] The same applies to the *superego* (see Kaplan-Solms & Solms, 2000).

and it continued to govern her behavior *unconsciously*. She had lost only a specific *aspect* of ego functioning.

Now let us consider Galin's hypothesis again, in relation to a second case, who lost a different aspect of language.

A patient who cannot express his thoughts in words

Although he was only in his 20s, Mr. J (see Kaplan-Solms & Solms, 2000, pp. 75–86), suffered a stroke (caused by bacterial endocarditis), mainly affecting Broca's area (see Figure 8.6). As a result, his speech lacked fluency, he spoke in a "telegrammatic" fashion, and he could only say a very few words (Broca's aphasia). His disability, which also included *hemiparesis* (i.e., paralysis of the right side of his body), had of course dramatically affected his life. He lost his job, his partner, and most of his

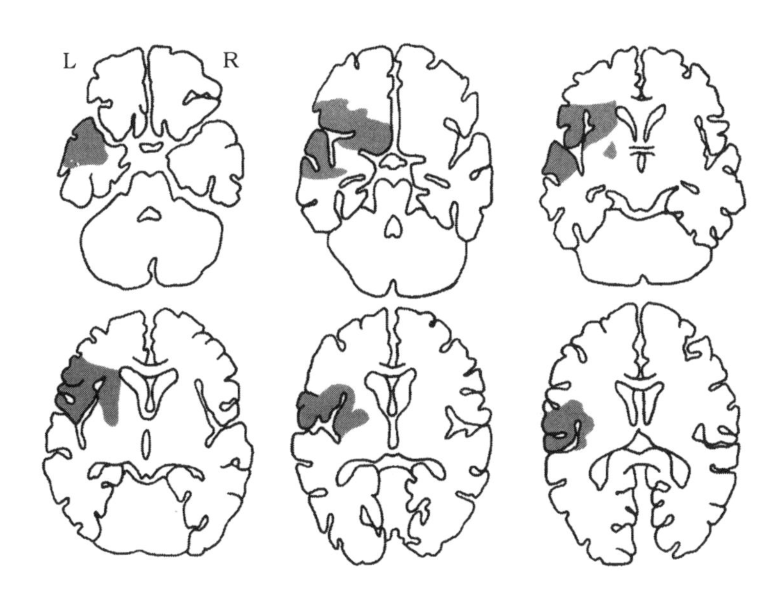

FIGURE 8.6
Lesion in patient with Broca's aphasia

friends. He understandably feared that he had no future prospects. All that he had previously taken for granted in life was slipping away from him. It was a tragic situation, and Mr. J was filled with anger, sadness, and loss.

When he was offered psychotherapy, he eagerly grasped the opportunity. There was much that he wanted to discuss, even though he no longer had the words to do so. One of the many things that he wanted to tell his therapist was that he now felt like "half a man." He communicated this by drawing a stick figure of a man, bisecting it vertically, and saying "man . . . halfie . . . halfie." This communication was pregnant with meaning. It conveyed the essence of his emotional situation, including its repressed ramifications, and it simultaneously linked them symbolically with his neurological (hemiparetic) condition. He had lost his masculinity and the self-esteem that was attendant upon it. However, he worked extremely hard in his psychotherapy to come to terms with these losses, and ultimately he was able to construct a new, viable life for himself, built on revised premises and priorities.

In short, this was a patient who was almost literally wordless, and yet he was able to make productive use of psychoanalytic therapy—the so-called talking cure—to negotiate the painful process of mourning and gain new insights about himself that enabled him to endure, with great courage, circumstances that would defeat many people with perfectly intact brains.

This case, no less than that of Mrs. K, does not support Galin's hypothesis. Notice that although this patient—like Mrs. K—sustained damage to the left hemisphere (the supposed seat of the system *Cs*), his consciousness and his executive ego were not impaired in any perceptible way. Clearly, the hypothesized metapsychological consequences of detaching "word presentations" from the unconscious "objects" or "things" that they represent (supposedly in the right hemisphere) occur only with damage to one small part of the left hemisphere—to the extent that they occur at all. The attachment of reflexive consciousness to thought processes is therefore only *one aspect* of left-hemisphere functioning. And the gross (executive) ego deficits predicted by Galin's hypothesis do not appear to materialize with left-hemisphere lesions at all.

Patients who cannot use language to regulate behavior

There *are* patients with focal neurological lesions who present with the sort of generalized ego deficits that Galin's hypothesis predicted. We briefly described a series of cases like this near the end of chapter 3. These patients' beliefs are riddled with contradictions, their perception of external reality is overwhelmed by their wishful phantasies, they appear to have no sense of time, and their thinking is grossly distorted by primary-process transformations. But these patients *have not suffered left-hemisphere damage* (see Figure 8.7). In fact, the lesions in these cases are

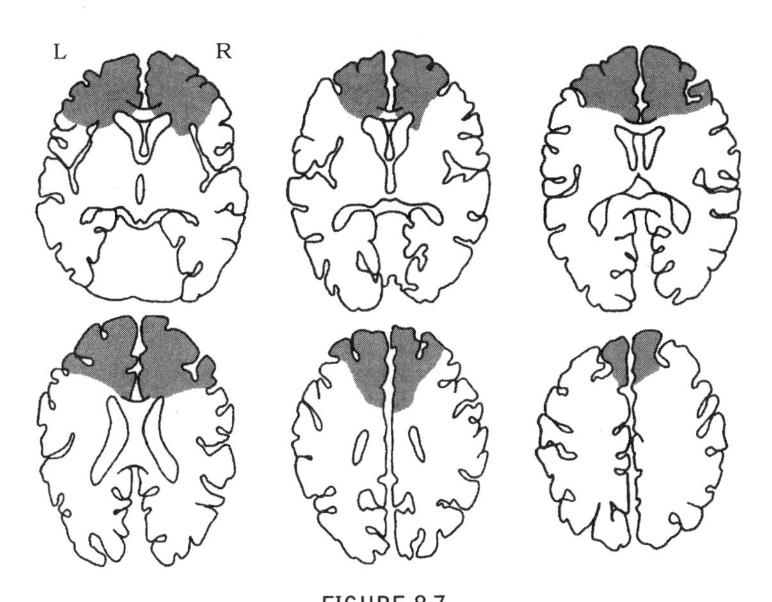

FIGURE 8.7
Lesion in patients with regulation deficits

not lateralized at all: they are typically *bilateral*. More important than which *side* of the brain is affected is the question of whether it is the front or the back of the brain that is involved. Only patients with *frontal-lobe* lesions present with such deficits.

Deep frontal-lobe lesions produce disorders of what Luria called the *regulatory* function of language. This aspect of language, also known as "inner speech," enables one to subordinate one's behavior to verbal programs, such as "first I must do this, before I can have that." We rely on this aspect of language all the time. It is easiest to recognize it in young children, who still often externalize their inner speech, thus making it clear how they are using words (often adopted wholesale from their parents) to regulate their behavior and impulses. Consider the common example of the small child who points at the thing that it desires, while simultaneously saying "No" or "Dangerous" to itself. With time, these self-instructions become increasingly internalized and invisible—that is, they become automatic and unconscious. These "regulatory" functions of language are lost with bilateral frontal-lobe lesions, especially those affecting the ventromesial frontal region. The distribution of the functions of the systems *Cs* and *Ucs* between the hemispheres is therefore no simple matter.

MORE ABOUT THE NEUROPSYCHOLOGY OF THE RIGHT HEMISPHERE

The right hemisphere is conventionally said to be specialized for *spatial* cognition (see above). Where damage to association cortex in the left hemisphere produces disorders of various aspects of language, damage to the equivalent parts of the right hemisphere produces disorders of spatial cognition. These patients cannot draw a bicycle without misaligning the component parts; they cannot copy a simple construction made with children's blocks; and they cannot learn the route from their bed to the toilet (for a detailed review see DeRenzi, 1982). However, some right-hemisphere functions do not sit easily under the heading "spatial cognition." This is readily apparent from the syndrome that most typically occurs with right parietal-lobe damage. This pattern of signs and symptoms (known as the "*right-hemisphere syndrome*") has three cardinal components. One of the components comprises the unequivocally "spatial" deficits just described (such as **constructional apraxia** and **topographical disorientation**), but the two other components of the syndrome are more complex. These go by the names **neglect** (or *hemispatial neglect* or *hemineglect*) and **anosognosia**.

Neglect

Patients with this condition neglect (i.e., ignore) the left-hand side of space (for a detailed review see Robertson & Marshall, 1993). If, for example, you stand to the right of such a patient and ask "How are you today Mrs. Jones?" she is likely to reply "Fine, thank you." If you stand to her left and ask the same question, she is likely to simply ignore you. This is not because she doesn't see or hear you. Neglect is a disorder of *attention* rather than perception. This problem affects not only objects in external space, but even the left half of the patient's *own body*. Such patients frequently shave only the right-hand side of the face, dress only the right-sided limbs, and eat only the food on the right-hand side of the plate.

Anosognosia

Anosognosia means unawareness of illness. When Mrs. Jones says she is "Fine, thank you," she really *means* it—even though (as a patient with a substantial right-hemisphere lesion) she is quite likely to be paralyzed down the left side of her body. Although they cannot walk and they need to use a wheelchair to get around, these patients claim to be fine and insist that there is nothing wrong with them. Their lack of awareness of their incapacities, and their rationalizations concerning their problems, extend to the point of delusion (for detailed examples see Ramachandran, 1994; Ramachandran & Blakslee, 1998; Turnbull, 1997). If, for example, a patient who claims that she is able to run is asked why she is in a wheelchair, she might respond: "There was nowhere else to sit." If asked why she is not moving her left arm, she could say something like: "I exercised it a lot earlier today, so I'm resting it." And so on. These patients seem prepared to believe anything, so long as it excludes admitting that they are ill.

This is literally true. Not uncommonly these patients make bizarre claims about their paralyzed limbs, such as denying that their paralyzed arm belongs to them and saying that it belongs to someone else (known as **somatoparaphrenic delusion**). They also frequently express intense dislike and hatred toward the paralyzed limb, beg surgeons to amputate it, and may even physically assault the limb themselves (known as **misoplegia**).

Milder cases suffer from **anosodiaphoria**. These latter patients do not frankly deny that they are ill, but they seem *indifferent* or *unconcerned* about it. They acknowledge their deficits intellectually but seem unaware emotionallly of the implications.

UNDERSTANDING THE RIGHT-HEMISPHERE SYNDROME

The range of symptoms just described cannot be reduced to "disorders of spatial cognition." Although there is a spatial component to these symptoms, some aspects of the right-hemisphere syndrome could just as well be described as "disorders of *emotional* cognition." The emotional functions of the right hemisphere are now generally recognized, and many aspects of the problem have been comprehensively studied. The same applies to the "attentional" functions of the right hemisphere.

Various theories have been advanced in recent years that attempt to account for the nonspatial aspects of the right-hemisphere syndrome. The first of these is the *attention-arousal hypothesis,* which accounts for neglect and the attentional aspects of anosognosia but for little else about the syndrome (see Heilman & van den Abell, 1980; Mesulam, 1981). According to this theory, the right hemisphere attends to both the left and the right sides of space, whereas the left hemisphere only attends to the right-hand side. Accordingly, when the left hemisphere is damaged, bilateral (right-hemisphere) attention is preserved, but when the right hemisphere is damaged, only unilateral (left-hemisphere) attention remains.

A second theory attempts to account for the emotional aspects of the syndrome but ignores the spatial aspects. This might be called the *negative-emotion hypothesis*. According to this theory, the right hemisphere is specialized for negative emotions, whereas the left is specialized for positive emotions. Damage to the left hemisphere thus reduces the capacity for positive emotion, causing depression (and so-called **catastrophic reactions**, which are more common with left- than right-hemisphere lesions), whereas damage to the right hemisphere has the opposite effect: the patient is inappropriately happy. Although this simple dichotomy between positive and negative emotions may seem rather ridiculous to our readers (cf. chapter 4), it is a surprisingly respectable and influential theory.

The *somatic monitoring hypothesis* (proposed by Damasio, 1994) is the third of the theories we review. This theory is based on the idea that the right hemisphere is specialized for somatic awareness (awareness of the body as a "thing"). Since, as we learned in chapter 4, emotion is generated—in part—by awareness of one's bodily state, right-hemisphere damage impairs emotional awareness. This theory is more sophisticated than the previous two, and it appears to accommodate all the major features of the right-hemisphere syndrome (spatial, emotional, and attentional), but we shall soon see that it, too, has problems.

It is interesting to note the simple reasoning behind all these theories. Initially, investigators observed that right-hemisphere lesions cause defects of spatial cognition, so they hypothesized that the right hemisphere might be specialized for spatial cognition. Then they observed that right-hemisphere lesions also cause defects of attention, so they added that the right hemisphere might be specialized for attention-arousal. Then they noticed that right-hemisphere patients are inappropriately unconcerned about their deficits, so they added that the right hemisphere may be specialized for negative emotions. Then, to account for the fact that right-hemisphere patients are unaware of the state of their own bodies, they hypothesized that the right hemisphere is specialized for somatic monitoring. This last hypothesis, as we have said, is more sophisticated than the others, but all of them are remarkably simplistic from a *psychological* point of view. The underlying reasoning is typical of the clinico-anatomical method (see chapter 2) and therefore not essentially different from Broca's: if something is clinically deficient due to brain damage, then the damaged tissue must have been specialized for producing that (now-deficient) thing. Psychotherapists have learned to mistrust this type of reasoning when it is applied to the *emotional* life of human beings. This is essentially due to the fact that psychotherapists (by and large) hold a *dynamic* conception of emotional life. They are therefore not surprised to find that the underlying mechanism of a disorder often turns out to be the very opposite of what it appears to be. A patient might appear to be inappropriately happy, not because he cannot generate negative emotions, but because he cannot *bear* them. An inappropriately happy person might, beneath the surface, very well be struggling with depression.

A PSYCHOANALYTIC PERSPECTIVE ON THE RIGHT-HEMISPHERE SYNDROME

The observation that right-hemisphere patients are inappropriately unconcerned is not based on deep psychological investigations. It is based on simple bedside evaluations of mood or on psychometric pencil-and-paper tests like the MMPI or Beck Depression Inventory, which rely on the patient's *own* assessment of their mood. Psychotherapists are rightly skeptical of theories based on measurements of this sort, but neuroscientists are in no position to know better. In this respect, they need to be educated.

Five patients with damage to the perisylvian convexity of the right hemisphere (Figure 8.8) were recently investigated in psychoanalytic psychotherapy. The purpose of this research was to

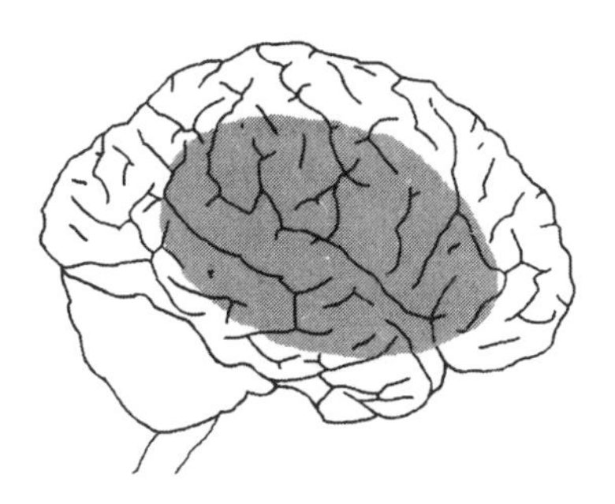

Figure 8.8
Right perisylvian region

investigate the underlying psycho*dynamics* of the personality changes caused by right-hemisphere damage (see Kaplan-Solms & Solms, 2000).

The first two patients exhibited typical features of the right-hemisphere syndrome—they were incompletely aware of their (substantial) cognitive and physical deficits, and they neglected the left-hand side of space (including the left side of their own bodies). They also displayed classical emotional indifference to their disabilities. However, this "indifference" was found to be a highly fragile state. In their psychotherapy sessions, both patients burst into tears for brief moments during which they seemed to be overwhelmed by emotions of the very kind that are normally conspicuous by their absence. This gave the impression of *suppressed* sadness, grief, dependency fears, and so on, rather than a true *absence* of such feelings.

For example, one of these patients—Mrs. B—found herself suddenly bursting into uncontrollable tears while reading a book (Kaplan-Solms & Solms, 2000, pp. 167–172). She then regained her composure and continued reading. When asked the next day by her therapist what she had been reading when she started to cry, she could not remember. All that she could recall was that it had something to do with a court case. On further investigation, it turned out that she had been reading about a court case involving parents who were fighting for compensation for a thalidomide child. Mrs. K, who had suffered a severe stroke during childbirth and lost the use of her left arm and leg, had clearly identified her own disability with that of the thalidomide child. However, she was completely unaware of this connection. The same patient (who was of Eastern European, Jewish descent) burst into tears repeatedly while watching the film *Fiddler on the Roof.* It would clearly be erroneous to claim that this patient could not *experience* negative emotions; more accurate would be to say that she could not *tolerate* them (particularly feelings of loss). Some empirical investigations with a patient of this sort show that such feelings of loss are the predominant affective feature (Turnbull & Owen, in press).

The second case was a man, Mr. C (Kaplan-Solms & Solms, 2000, pp. 160–167). He too was paralyzed by a right-hemisphere stroke but was "unaware" of his deficit. Accordingly, his physiotherapist was unable to enlist his cooperation in trying to teach him how to walk again. He seemed oblivious of his deficit and totally unconcerned about it. When recounting the relevant events to his psychotherapist the next day, however, he suddenly burst into tears. When she probed the associated feelings, Mr. C blurted out: "But *look* at my arm—what am I going to do if it doesn't recover? How am I ever going to work again?" He then regained his composure and reverted to his typical "indifferent" state. (This behavior is not consistent with the somatic monitoring hypothesis. Mr. C was not unaware of the state of his body. Rather, he had suppressed *conscious* awareness of the state of his body. Attention is not an emotionally neutral function.) As with Mrs. B, such occurrences were common with this patient. They were also not very difficult to understand. Both of these cases were intolerant of the depressive feelings associated with their loss (which they were certainly *unconsciously* aware of), and they were therefore unable to work through these feelings by the normal process of *mourning*.

MOURNING AND MELANCHOLIA

Failures in the process of mourning take many forms. In a famous paper called "Mourning and Melancholia," Freud (1917e [1915]) contrasted the normal process of mourning with the pathology of melancholia (i.e., clinical depression). He argued that, in mourning, a person gradually comes to terms with loss by giving up (separating from) the lost love object, whereas in depression this cannot happen because the patient *denies* the loss. You cannot come to terms with a loss if you do not acknowledge that it has happened. Freud said that this was particularly apt to occur if the original attachment to the lost object had been a *narcissistic* one. (In a narcissistic attachment, the separateness of the love object is not recognized, and it is treated as if it were part of the self. Narcissism is contrasted with *object love*, which is a more mature form of attachment, where the independence of the love object is acknowledged.) Freud showed that in melancholia, the patient denies the loss of the love object by identifying himself with it (by literally becoming that object in phantasy). The depression itself then results from the internalization of the feelings of resentment toward the object that has abandoned him. (The narcissist attacks the internalized object with all the ruthless vengefulness of a lover scorned.)

This explanation seems to hold good for the third case of right-hemisphere syndrome that was investigated psychoanalytically. This case—Mrs. A (Kaplan-Solms & Solms, 2000, pp. 173–179)—suffered severe spatial deficits, neglect, and anosognosia, but at the same time she was profoundly depressed. This is unusual for right-hemisphere patients, producing a paradoxical situation in which the patient was unaware of a loss (anosognosia) and yet simultaneously displaying severe depressive reactions to it. She was constantly in tears, lamenting the fact that she was such a burden to the medical and nursing staff, whose generous attention she did not deserve since she was not fit to live, and so on. The psychoanalytic investigation revealed that Mrs. A was, in fact, unconsciously very much aware of her loss, but she was denying it by means of the introjective process described by Freud. Unconsciously, Mrs. A *did* have an internalized image of her damaged, crippled self, and she attacked that image to the point of twice attempting to kill herself.

In this case, the patient was *overwhelmed* by feelings of the same type that the previous two patients managed (for the most part) successfully to suppress. In two final cases, the situation was more complicated still.

Defenses against melancholia

The two patients—Mr. D and Mr. E (Kaplan-Solms & Solms, 2000, pp. 187–197)—were anything but unconcerned and indifferent about their deficits: they were absolutely obsessed by them. They also displayed a symptom mentioned earlier: *misoplegia* (hatred of the paretic limb). One of these patients (Mr. D) had only a mild paresis of the left hand, and he would have been able to use it if he had tried. However, he refused to use the hand, and he actually demanded that the surgeon cut it off because he loathed it so much. Mr. D once became so enraged at his hand that he smashed it against a radiator, claiming that he was going to break it to pieces and post the bits of flesh in an envelope to the neurosurgeon who had operated on him. This conveys vividly the emotional state of such patients.

It is interesting that *the same lesion site* can produce such opposite emotional reactions: unawareness of a limb and denial of its deficits, versus obsessive hatred of a limb and its imperfections. This state of affairs almost *demands* a psychodynamic explanation. The psychoanalyst who treated these two patients came to the conclusion that their underlying psychodynamics were very similar to those of Mrs. A: they, too, attacked their internal awareness of their loss, but rather than attempt to kill themselves (like Mrs. A), they reacted by trying literally to detach the hated (damaged) image of themselves—or parts of themselves—from the rest of themselves, in order to preserve their intact selves.

No doubt, other permutations are possible.[10] What all of these cases have in common is a failure of the process of mourning. Underlying the range of clinical presentations was this common dynamic mechanism: these patients could not tolerate the difficult feelings associated with coming to terms with loss. The superficial differences between the patients are attributable to the fact that they defended themselves against this intolerable situation in various ways.

The reason mourning fails in these patients

We are now in a position to integrate these findings. The right perisylvian convexity is specialized for spatial cognition. Damage to this area therefore undermines the patients' ability to represent the relationship between self and objects accurately. This in turn undermines object relationships in the psychoanalytic sense: object love (based on a realistic conception of the separateness between self and object) collapses, and the patients' object relationships regress to the level of narcissism. This results in narcissistic defenses against object loss, rendering these patients incapable of normal mourning. They deny their loss and all the feelings (and even *external* perceptions) associated with it, using a variety of defenses to shore up their denial whenever the intolerable reality threatens to break through.

Left-hemisphere patients, by contrast, retain the capacity for object love, for the reason that the requisite "spatial" concepts remain intact. Accordingly, these patients, whose objective loss is at least equivalent to that of right-hemisphere patients, *are*

[10] Moss and Turnbull (1996) described a 10-year-old child, with the classic right-hemisphere syndrome, who alternated between a state of denial (anosognosia) and hatred (misoplegia) in relation to his left hand. During the period when he hated it, he said that he wanted to have that arm surgically removed and replaced with the left arm of his mother.

able to negotiate the difficult process of mourning. The "depression" and so-called catastrophic reactions of left-hemisphere patients are, in fact, healthy and appropriate responses to devastating loss. Right-hemisphere patients, however, stuck in their narcissism, cannot test their phantastic misconceptions against the perceived reality (as Mrs. K did), and they cannot undertake the normal work of mourning that Mr. J did.

Psychoanalytic investigation of the inner life of neurological patients clearly has much to offer us. In this instance, it was able to throw important light on a syndrome that was inadequately accounted for by a variety of neurocognitive theories, each of which failed to accommodate the psychological complexities of human emotional life. But psychoanalytic hypotheses are no less prone to error than cognitive ones, and they therefore need to be subjected to the same rigorous empirical tests.

THE RIGHT HEMISPHERE AND THE EGO

Galin's hypothesis is a good example of the need for rigorous testing. The hypothesis predicted that left-hemisphere patients should have gross ego deficits of psychotic proportions (which they do not). In contrast, right-hemisphere patients, being deficient in respect of primary processes, should be *more* realistic and rational than are normal people (which they, too, are not). If anything, the right-hemisphere patients displayed *greater* ego deficits (more primary-process thinking) than did their left-hemisphere counterparts. Still greater disturbances in this direction were observed in cases with *bilateral* damage in the ventromesial frontal region—a part of the brain that is minimally asymmetrical.

Galin's hypothesis is clearly unsustainable. The left hemisphere is not the seat of Freud's system *Cs*, and the right hemisphere is not the seat of the *Ucs*. Both cerebral hemispheres subserve different *aspects* of ego functioning. Moreover, the aspect of ego functioning that Galin most wanted to attribute to the left perisylvian convexity—the inhibitory function of the secondary process—is more appropriately linked with the ventromesial region of *both* frontal lobes (see chapter 3). For its part, the primary processes of the unconscious id appear to have more to do with the primitive "state-dependent" influences that the subcortical structures we reviewed in chapter 4 bring to bear on the cortical mantle than with the "channel-dependent" information-processing functions of the cerebral cortex itself (see chapter 1).

CONCLUDING REMARKS

In this chapter, alongside our brief review of the neuropsychology of cerebral hemispheric asymmetry, we have attempted to introduce two additional themes that are the main focus of the final chapters of this book. These are: (1) our belief that psychoanalytic hypotheses concerning the functional organization of the human mental apparatus *can* (*and should*) *be tested* using modern neuroscientific methods, and (2) our conviction that, as neuroscientists begin to tackle the complexities of the inner world, they have much to gain from interdisciplinary collaboration of this sort.

CHAPTER 9

THE SELF AND THE NEUROBIOLOGY OF THE "TALKING CURE"

Lest the reader be disappointed, it seems fair to admit at the outset that we are not yet in a position to give a proper account of either the "self" or the "talking cure" in neuroscientific terms. But we do have some tantalizing clues, and that is reason enough to consider these issues briefly here, even if only to clarify what research still needs to be done. We begin by reviewing the material we have covered already, and pulling together some of the main strands, to try to develop a coherent overall picture of how the mind works.

HOW THE MIND WORKS, SUMMARIZED

In chapter 2, we suggested that the mental apparatus is knowable in two different ways. By looking inward, we gain a subjective impression of our minds—a view from the inside, as it were. This is the method of studying the mind that psychoanalysis uses. The physical organ of the brain provides a second perceptual viewpoint on the mind—an "objective" perspective—a view of the mind as a thing; this is what the mind looks like when it is viewed from the outside. The fact that the mind can be viewed in these two different ways is the basis of the mind–body problem—the *illusion* that the mental apparatus consists of two different kinds of "stuff."

This illusion can be turned to our advantage, because it allows us to check conclusions that we reached from one point of view against those that we reach from another. As in the famous allegory of the blind men and the elephant, limiting oneself to information gained from a single perspective can produce misleading results. If we "blind men"—and women—of science cooperate, we will gain a more accurate picture of what the mind actually is. It is for this reason that psychoanalysis, at this point in its history, has a great deal to gain from collaboration with the neurosciences, and vice versa. The topics making up the chapters of this book were partly selected to demonstrate this point. Questions like "What is feeling? What is consciousness? What is the self?" cut to the very heart of mental life.

Amazingly, these were not the sorts of questions that bothered neuropsychologists and behavioral neurologists during the century just past. The subjects that preoccupied behavioral neuroscientists then were far removed from these issues, which are of such fundamental concern to most ordinary human beings. For many years, neuroscientists studied the elemental conditions of perception and movement. To the extent that they ventured into the interior of the mind at all, they did not stray very far beyond the behavioral (read: externally observable) surface, focusing on language, memory, and problem solving. They shied away from the real stuff of the inner world, such as feelings, consciousness, and the self. Happily, all of this changed (quite suddenly, it seems) in the last years of the twentieth century, and now we are already in a position to ask such daring questions as "How does the talking cure work?"

To begin to answer that question, we need first of all to remind ourselves of a conclusion that we came to in chapter 2, when we asked what the defining characteristic of the "mind" was. We concluded that the essence of the mind is conscious *awareness*. How, then, does awareness work?

THE CORE OF AWARENESS

The brain is an organ that promotes our survival as biological creatures. It does this by mediating between the inner needs of our bodies and the dangers and delights of the outside world—the location of all the objects that satisfy our inner needs.

The brainstem is the anatomical core of the brain and, in evolutionary terms, its oldest part. Within the brainstem there are a number of nuclei that regulate our vegetative, visceral life. They control heartbeat, respiration, digestion, and the like. The design of these circuits is "hard-wired," and the basic design is shared by all mammals. These circuits are so crucial to life that if there were to be even minor variations in their structure and connectivity, we would not survive. They have been preserved so long through evolutionary history precisely because they work so well. While this is a fascinating brain region if you are a neurologist, these circuits have little directly to do with the mind—whose business it is to mediate *between* such things and the perceptual-motor world outside.

The mind begins where these systems end. Just above these circuits, in the upper part of the brainstem, lies a set of structures that participate in the regulation of visceral *as well* as mental life, in one particular way (see chapter 3). They govern the activational tone (or "state") of the brain, *which we perceive subjectively as the background medium of our conscious awareness*—the "page" onto which the ever-changing contents of perception (and thought) are written. This page is never really blank, even during sleep.

The inner source of consciousness reflects the current state of our bodies. To be more precise, it reflects the current state of our inner needs. This infuses the background "tone" of conscious awareness with a particular *quality* of feeling. The inner surface of consciousness, if its tone were to be conveyed in words (which it is not), would say something like: "I exist, I am alive, and I feel like *this*."

EXTERNAL SOURCES OF AWARENESS

The other aspect of "core consciousness" (to use Damasio's term) derives from the world around us. The stimuli that inscribe representational "contents" onto the page of consciousness are registered primarily in the posterior part of the forebrain, in a series of structures dedicated to the reception, analysis, and storage of information about the world. These structures combine myriad stimuli arriving from our various sense organs into the recognizable "objects" that constitute the physical world as we know it. A unit of consciousness—a moment of awareness—consists in a coupling of these two things: a momentary state of the core self in relation to its concurrent surround of objects. The essence of consciousness is therefore a *relationship*: "I feel like *this* in relation to *that*." This relationship reflects the fact that our inner needs can only be satisfied by things that exist beyond ourselves. Our feelings (the inner sources of consciousness) are therefore always defined in relation to the objects of our needs (the outer sources of consciousness).

It does not require a great leap of imagination to see why things are arranged like this. This is what consciousness is *for*. It tells us how we *feel* about things. The things in question are, basically, things like: "I feel hungry, I want one of those; I feel sexually aroused, perhaps she/he will oblige; I feel scared, I think I'll get out of here." If consciousness did not require *feelings*, we would get along fine without it. The reception, analysis, and storage of information—and the programming, regulation, and verification of action—do not depend on consciousness. Computers can (and do) perform such functions, and *we* perform most of them *unconsciously* most of the time (see chapters 2 and 3). Conscious awareness, in its essence, then, imparts *value.*

INHERITED MEMORIES: THE BASIC EMOTIONS

Built on the foundations of core consciousness, both conceptually and anatomically speaking, is a set of connections that encode self–object relationships of universal significance. These are connections that link certain feeling states with certain classes of perception, which in turn, when activated, trigger "pre-prepared" motor programs. These connections are the "*basic-emotion command systems*" (as Panksepp calls them). They enable us to respond *automatically*—in ways that promote survival—to events of biological import. We do not need to learn these response patterns ourselves: they are transcribed in our genes by virtue of the fact that they promoted the survival of countless generations of our ancestors, over aeons. This precious biological legacy represents the core value system of our species (and, indeed, of all mammals).

There are four of these basic-emotion command systems: the SEEKING (and associated pleasure-lust) system, the RAGE (or anger-rage) system, the FEAR (or fear-anxiety) system, and the PANIC (separation-distress) system. This classification does not derive from Plato, nor from any other philosopher. It flows from many lines of converging evidence in neuroscience, from the painstaking observation of nature.

There are constitutional variations in the relative sensitivity of these emotion command systems in different individuals. The systems are standardized and run over highly predictable fiber pathways, employing specific chemical messenger systems. But the principle of individual variation across the population applies to these systems no less than to any other part of the body. Moreover, as we learned in chapter 7, these systems, like other parts of the brain, mature in each individual in a particular environmental context. This context—particularly during certain "critical periods"—fills in the many "blanks" that necessarily exist in such systems. We know— because our ancestors learned this for us—that anything that causes you to feel pain is best avoided. However, the "things" that will actually impart this pain need to be discovered afresh by each one of us (witness the behavior of any baby or toddler in this respect.) There are many potentially noxious objects in every environment that evolution could never predict (electric sockets, for example). For these reasons, although we all have four basic-emotion command systems, rather than three or five, and although they do roughly the same things in all of us, nevertheless each of us still has to develop his or her own individual classification of the "good" and "bad" objects in the world.[1] In this way, through complex interactions between our genes and the maturational environment, we develop a personal version of the world—an *inner world*—that is uniquely our own.

EXTENDED CONSCIOUSNESS

As reviewed in chapter 5, much of human memory is unconscious, and it never becomes conscious—though that does not mean that it does not *influence* consciousness. Most of what we

[1] Of course, this does not mean that *none* of the "contents" of the basic-emotion command systems is prewired. It also does not imply that there is little overlap between our individualized classifications of the world. The basic structure of the classification is determined by our shared, biologically driven value systems, and much of what we experience (particularly at a given moment of history, in a given culture) is not unique to the individual.

do consciously, in our moment-to-moment lives, depends upon implicit (unconscious) memory systems, which exert their effects on us without us even realizing it. Our every conscious moment is shaped by unconscious events, derived from a personal and biological past of which we usually have no inkling. "Inherited" memories determine the *form* of the basic-emotion command systems. The "good" and "bad" objects mentioned above determine the *contents* of those systems. This is the core of a system of implicit learning, which we reviewed in chapter 5. Some people (Joseph LeDoux, for example) call the motivational core of this type of memory system "emotional memory."

These are unconscious influences on consciousness, derived from the past. Consciousness itself is extended beyond the immediate present by our capacity to "replay" interactions with objects (good, bad, and indifferent) in our mind's eye. This is explicit (conscious) remembering, the most important variety of which (from our subjective point of view) is called "episodic memory"—memory for "personal" events.[2] Episodic memory supplements our immediate experience of core consciousness (self–object couplings derived from current perception) with reminiscences of past moments of consciousness (past self–object couplings). The reactivation of such couplings of the self (inner consciousness) with stored information derived from past events (outer consciousness) seems to be the task, above all, of the hippocampus. Episodic memory links traces of past events (registered primarily in the posterior cortical networks) with the fact that you were there and *felt* something. This is what makes it familiar, and familiarity is the core of episodic memory. (The sense of familiarity is also fallible—hence the *déjà vu* phenomenon—and perhaps also the problem of "false memories.")

[2]The other main variety of explicit memory is, as discussed in chapter 5, "semantic memory"—memory of *facts* as opposed to *events*.

THE IMPORTANCE OF THE EXECUTIVE

We come now to the nub of the subject matter of this chapter. Everything described up until this point implies a more or less *passive* mechanism. But the most important distinguishing characteristic between the inner and outer perspectives on the mind is the experience of an active *agency*. This sense of agency is synonymous with the sense of *self*. The self can only be perceived subjectively. When the mind is observed externally, as a physical object, then the *agent* of the mind is literally invisible. But the external perspective allows us to study its physical correlates objectively, and this helps to throw its functional organization into relief.

At its lowest levels of organization (at the level of core consciousness), the primal SELF (to use Panksepp's terminology again) is a brainstem structure (see Figure 4.2). It is essentially the inner source of awareness described above: the source of the experience of "being alive." But it is a mistake to think of this source of consciousness in purely sensory terms. Although it is true that the inner surface of core consciousness *perceives* the current state of the body, it is nevertheless, fundamentally, a *motor* system. There are two reasons for this, mentioned repeatedly in previous chapters: first, the sole purpose of perception is the guidance of *action*; second, the fundamental purpose of consciousness is the perception of *emotion*. That is to say, the SELF guides action on the basis of *evaluation*.

At the level of organization of the basic-emotion command systems, such evaluations have an *obligatory* outcome. They trigger stereotyped motor programs (reflexes and instinctual behaviors). Such reactions are compulsive. At this primitive level of organization, therefore, the self is still essentially a passive mechanism. Although it triggers motor programs, it lacks *choice*. It is dominated by what Freud called the "repetition-compulsion." In short, this primitive self is devoid of *free will*.

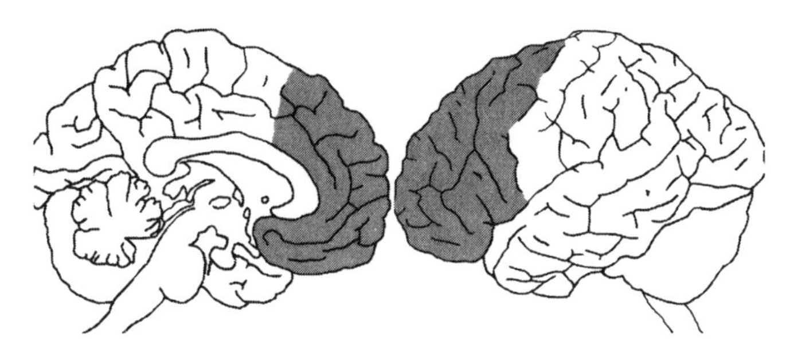

FIGURE 9.1
The prefrontal lobes

From the neuroscientific point of view, ironically perhaps, the essence of "free will" appears to be the capacity for *inhibition*—the capacity to choose *not* to do something. What distinguishes human beings more than anything else from their nearest primate relatives is the development of a higher-level "self" system, which is organized fundamentally on *inhibitory* mechanisms. These mechanisms, which have their physical locus in the **prefrontal lobes** (the crowning glory of the human brain), bestow on us the capacity to *suppress* the primitive, stereotyped compulsions that are encoded in our inherited and emotional memory systems. On this basis, the inhibitory prefrontal lobes may be regarded, with some justification, as the very tissue of our humanity (Figure 9.1).

HOW DOES THE EXECUTIVE SYSTEM WORK?

Although it is sometimes better for decisions about action to be taken rapidly without your having to think about them, the frontal lobes offer the potential to *delay* (inhibit) such decisions in the interests of *thinking*. Thinking may be regarded as *imaginary* acting, whereby the outcome of a *potential* action is *evaluated*. This is achieved by running the envisaged action programs while motor output is precluded (inhibited). Acting without acting is thinking (imaginary acting). Inhibition is therefore the prerequisite *and* the medium of thought.

The prefrontal lobes mature after birth, mainly in two massive spurts around the ages of 2 and 5 years, but they continue to develop throughout the first two decades of life. They are therefore heavily "experience-dependent." The experiences that shape the activity of these executive mechanisms in the earliest years of life will determine their individual structure. The application of their inherent (neurochemical) inhibitory capacities is, accordingly, literally *sculpted* by the parental (and other authority) figures who guide this aspect of the child's development during the critical early years.[3] This "sculpting" process appears to be governed by at least two things: first, by what parents *do*; second, by what they *say*.

Mirror neurons

"**Mirror neurons**" are located on the outer surface of the frontal (and parietal) lobes (Gallese et al., 1996; Rizzolatti & Arbib, 1998; Rizzolatti et al., 1999). Their mode of working (discovered in monkeys) is best captured by the phrase "Monkey see, monkey do" (Carey, 1996). When a monkey *does* something, the neurons in its motor cortex fire in the characteristic pattern that shapes the behavior in question. What Rizzolatti's group discovered is that the motor neurons in a second monkey, who is only *passively observing* the behavior of the first monkey, *fire in the same pattern* as those of the first monkey—thereby mirroring the observed behavior "in imagination." This class of neurons has thus far only been observed in (cortical) action systems. When

[3] See Schore (1994) for a comprehensive review of many relevant aspects of this complex developmental process.

neuroscientists are bold enough to look for them in the core emotion systems, we hazard a guess that they will form the basis for a neurobiology of empathy. Although the existence of this mechanism in human children has yet to be demonstrated directly, it seems reasonable to assume that this is the physiological mechanism whereby children "internalize" the behavior of their parents. In this way, executive programs are established by repeated activation through *observation*, without the relevant behaviors having to be actualized. Passivity is thereby transformed into activity (e.g., self-restraint) at the same time as action is transformed into thought.

Inner speech

Children also internalize what their parents *say* to them, using the mechanism of "inner speech" described in chapter 8. They thereby transform *prohibitions* into *inhibitions*. Language is an enormously powerful tool of self-regulation. This is perhaps best demonstrated in the negative, by way of a pathological phenomenon known as "**dissociation between knowing and doing**," observed in patients with frontal-lobe lesions. The patient is asked to do something—for example, to stand up. The patient replies, "Okay," but fails to stand up. The examiner asks the patient, "What did I ask you to do?." The patient replies, "Stand up" (thereby demonstrating understanding of the instruction). "So what are you going to do?" the examiner asks. "Stand up," replies the patient (thereby demonstrating the intention to comply with the instruction), but still nothing happens. This is due to a loss of the *ability to regulate behavior verbally*. These patients can repeat an audioverbal program, but they cannot use the program to govern their behavior. This phenomenon only occurs in patients with (substantial) damage to the frontal lobes.

Reversing this pathological mechanism, we infer that the subordination of behavior to internalized verbal instructions is (at least in large part) a function of the frontal lobes. Initially, the child literally repeats what it hears, but gradually this behavior too is internalized, and once again action is transformed into thought (for a fascinating account of this developmental process see Luria & Yudovich, 1971).

Thinking consists, therefore, in at least these two forms of inhibited (or imaginary) action.

WHAT IS PSYCHOPATHOLOGY?

We have just reviewed two ways in which the frontal lobes gain control over the motor apparatus of the brain, and we have shown how these control mechanisms develop (during critical periods) through the internalization of parental words and deeds. These are the sorts of mechanisms that the mature self uses to suppress the automatic motor stereotypes reviewed previously (and in chapter 4). This is how it interposes *thought* between impulse and action.

There are, of course, many ways that this process can go wrong. Anything that undermines the capacity of the self to govern its action systems in an efficient way will constitute some form of psychopathology. The two simplest examples would be constitutional excesses of impulse or deficiencies of inhibition. However, when one considers the innumerable permutations introduced by the variety of "impulse" systems (basic-emotion command systems), combined with the almost infinite range of environmental factors that might affect their maturational processes as well as the maturation of the inhibitory systems that govern them, it becomes clear why there are so very many different things that go under the heading of "psychopathology."

Certainly, some of the basics are becoming clear. We now understand the various emotion systems that are "poorly regulated" in various psychological disorders. For example, in chapter 6 we discussed the likely role of the SEEKING system in (the positive symptoms of) schizophrenia. This system probably also has a role to play in bipolar mood disorder (Panksepp, 1998). The biological basis of the anxiety disorders has much to do with the FEAR system (LeDoux, 1996; Panksepp, 1998), though there is likely to be a key role for the PANIC (separation-distress) system as well. The biology of this latter system may also have much to do with autism and Asperger's syndrome, and probably with depression (Panksepp, 1998). This is not to suggest that these disorders have an entirely hereditary or "genetic" basis. This conclusion does not follow *at all* from the identification of the biological basis of these disorders. Indeed, it is probably the *regulation* of these systems, sculpted by cortical structures in the course of development, that determines whether (and which) disorders develop. This brings us back to the prefrontal lobes.

THE METAPSYCHOLOGY OF THE "TALKING CURE"

Psychoanalytically informed readers will recognize, by another name, the functional entity that we are discussing here. Freud attributed all of the functions that we have reviewed in this chapter to a metapsychological entity called the "ego." Not surprisingly, therefore, he conceived of the "talking cure" as a means of *strengthening* this ego—that is to say, of extending its sphere of influence over the two domains between which it is interposed: on the one hand, the "id" (roughly equivalent to the basic-emotion command systems); on the other, reality (which it controls via the motor systems). We discussed these executive functions of the ego in chapter 3 and linked them with the frontal lobes. As we said there, it is important to remember that

Freud *abandoned* his original idea that the functions of the mind should be divided between the systems Conscious (or Conscious and Preconscious) and Unconscious. In 1923, he recognized that the rational, reality-constrained, executive part of the mind is not necessarily conscious, and not even necessarily *capable* of becoming conscious (Freud, 1923b). Consciousness, for Freud, was therefore not a fundamental organizing principle of the functional architecture of the mind. Accordingly, from 1923 onward, Freud redrew his map of the mind (see Figure 3.3) and attributed the functional properties previously attributed to the "system *Cs-Pcs*" to the "ego"—where only a small portion of the ego's activities were conscious (or capable of consciousness). The ego was mainly unconscious. Its core functional property was the capacity, not for consciousness, but, rather, for *inhibition.* Freud considered this capacity (the capacity to inhibit drive energies) to be the basis of all the ego's rational, reality-constrained and executive functions. This inhibitory capacity was the basis of what Freud called "secondary-process" thinking, which he contrasted with the unconstrained mental activity that characterized the "primary process." It was this property (rather than consciousness) that gave Freud's ego—the "autobiographical self" of Damasio—executive control over the otherwise automatic, biologically determined functions of the mind.

How, then, does the talking cure "strengthen the ego"? According to Freud (1940a [1938]), it does so by reversing the process of "repression."[4] Repression involves the exclusion of portions of the mind from the functional sphere of influence

[4] Note especially that other forms of ego weakness—such as those attributable to "disavowal" rather than "repression" (resulting in psychotic rather than neurotic illness)—are, according to Freud, not treatable by the "talking cure." The talking cure is designed only to undo repressions (see Freud, 1924b [1923], 1940a [1938]).

of the ego. "The repressed" is exempt from the inhibitory constraints imposed by the "secondary process," and it therefore functions according to the compulsive, stereotyped "primary-process" mode of the id (or system *Ucs*—see above). The aim of the talking cure, then, is to bring to bear on the repressed the inhibitory constraints of the secondary process and, thereby, to bring them under the flexible control of the ego (the "self," or "free will").

Armed with the knowledge we have gained regarding the functional anatomy of the mental apparatus, it is not difficult to translate all of this into neuroscientific terms.

THE NEUROBIOLOGY OF THE "TALKING CURE"

We learned in chapter 1 that the prefrontal lobes form a superstructure over all the other parts of the brain. This gives them the capacity to integrate all the information streaming into the brain (from its current visceral and environmental situation) with all the information derived from previous experience stored elsewhere in the brain—and then *to calculate the best course of action* before executing a motor response.

"Repression" may therefore be defined as anything that short-circuits this process. Any part of the brain's activity that is excluded from the overarching network of executive control exercised by the prefrontal lobes is, in a sense, the repressed. This, in turn, implies that there must be several varieties of repression. Indeed, we have implicitly come to this conclusion already, in our discussions (in previous chapters—e.g., in chapters 5 and 8) of *various* mechanisms all of which seemed to deserve the name "repression."

The aim of the talking cure, then, from the neurobiological point of view, must be *to extend the functional sphere of influence of the prefrontal lobes.*[5] It is therefore a matter of no small interest to us to learn that the few functional-imaging studies that have, to date, been conducted on the effects of (various forms of) psychotherapy, all show essentially the same thing (e.g., see Bakker, Van Balkom, & Van Dyck, 2001; Baxter et al., 1992; Brody et al., 1998; Ferng et al., 1992; Schwartz et al., 1996). In the first place, they show that the functional activity of the brain is indeed *altered* by psychotherapy. Second, they show that *specific* changes are correlated with the therapeutic outcome. Third, and most pertinently, they show that *these outcome-specific changes are essentially localized to the prefrontal lobes.*

HOW DOES THE TALKING CURE WORK?

How do these changes happen? How does the talking cure "extend the functional sphere of influence of the prefrontal lobes"? What we have said already implies that it probably does so in at least two ways. First, as its name implies, it uses *language*, which (as we learned in chapters 3 and 8) is an extremely powerful tool for establishing supraordinate, reflexive, and abstract connections between the concrete elements of perception and memory, and for thereby subordinating behavior to selective programs of activity. Second, it uses *internalization*, the mutative power of

[5] Needless to say, this aim must remain an ideal. It hardly seems possible—or even desirable—to bring every aspect of the brain's functional activities under the selective control of the prefrontal lobes. A large part of what goes on inside us will forever remain hidden from us. In this respect, it is important to remind readers once more that the functional domain of the ego is not synonymous with the functional domain of *consciousness*. Consciousness is an extremely limited entity (see chapter 3). The unconscious and the primary process are there for a *reason*. The *delay* involved in thinking is not always desirable and can even be positively dangerous (see chapter 4).

which (as we learned above) is probably largely confined to certain critical periods of frontal-lobe development (in the first few years of life), but which may perhaps be artificially rekindled by the regressive nature of the "transference" relationship. Of this we know virtually nothing, and the same applies to a great deal else of relevance to the question we have posed for ourselves.

The future will reveal the extent to which the processes we have hypothesized here *actually* determine the outcome of a psychoanalytic treatment. We must also leave to future research a host of other, more specific questions—such as whether different mechanisms are effective for different psychopathologies, which types of psychopathology are most amenable to talking therapy, and which aspects of the talking cure are most relevant in the different types.

It is appropriate that this short chapter should end on a note of uncertainty. This is not the place to reiterate the false certainties of the past concerning the therapeutic mode of action of psychoanalysis. The purpose of correlating our existing psychological insights (limited as they are) concerning the mode of action of the "talking cure" is not to shore up questionable theories but, rather, to reestablish our knowledge on a new, and more secure, scientific footing. With this goal in mind, then, we come to the final chapter of our book.

CHAPTER 10

THE FUTURE AND NEURO-PSYCHOANALYSIS

"Metapsychology" is an obscure term to the modern scientific ear. But metapsychology is ultimately what all of mental science is about, including cognitive neuroscience. Metapsychology is an attempt to describe the functional architecture of the mental apparatus (the instrument of our mental life) and to define the laws that govern its workings (see chapter 2). Functional architectures are abstractions—*virtual* entities. They are not things that can be directly perceived. They are *inferred* from the data of observation.

In chapter 3, we described the functional architecture of *consciousness* and the laws that govern its workings (insofar as we currently comprehend them). In chapter 4, we did the same for *emotion*; in chapter 5, for *memory*; and so on. All of these things are abstractions. You cannot perceive a "memory system." You can see the anatomical tissues *between* which the system is distributed, and you can experience *an individual reminiscence*, but the memory *system* itself is an abstraction. Cognitive neuroscience is ultimately about such things—"memory systems," "consciousness systems," "emotion systems," and so on. That is why we say that metapsychology (describing the functional architecture of the mental apparatus) is what cognitive science is really about.

In this respect, cognitive science is no different from any other branch of science. Physics, for example (and what could be more concrete than physics?), is about abstractions like "gravity," "electricity," "weak forces," and the like. These things too cannot be *seen*, but they are what physics is *about*. Each branch of science studies a different aspect of nature and aims to discover the laws that govern it; and such laws—the yield of science—always take the form of abstractions. They are virtual entities, inferred from the multiplicity of concrete things and events through which they are realized.

The mind is an aspect of nature like any other, and the "mental apparatus" is the abstraction that lies behind it, which we infer from our observations and aim to master scientifically. But the mental apparatus has one unique attribute that distinguishes it from other parts of nature: *It is the part of nature that we ourselves occupy*. It is *us*. This has the implication, not only that it matters more to us than any other part of nature, but also that we have a unique observational perspective on it. We know what it *feels* like to *be* a mental apparatus. We do not know what it feels like to be anything else.

For this unique reason, the memory and other mental "systems" that neuroscientists infer from their observations of neural tissues, under various conditions, can also be studied from the point of view of what it feels like to *be* such systems. We therefore have two points of view on all the different systems that comprise the mental apparatus—and, indeed, on the mental apparatus as a whole.

As we have said previously, this should be an asset to our science. And indeed it is. But historically, at least until now, we have failed to recognize this fact. Instead, we have behaved as if each of our two perspectives on the mental apparatus were studying a different piece of nature, and we have accordingly divided the mental apparatus into two kinds of "stuff," one of which (the brain) is studied "objectively" by neuroscientists, and the other (the self) is studied by psychoanalysis—the science of subjectivity (see chapter 2).

This misleading dichotomy is what this book has attempted to redress. We have attempted to show that when cognitive neuroscientists study the memory (and other) systems of the mental apparatus, they are studying the *same thing* that Freud was studying, and attempting to describe and define in his metapsychological writings—and to depict in his diagrams, such as the famous one that appeared in chapter 7 of *The Interpretation of Dreams* (Freud, 1900a, p. 538) (Figure 10.1). Such diagrams are identical in conception and purpose to the information-processing diagrams of contemporary cognitive neuroscience, such as, for example, neurophysiologist Allan Hobson's latest attempt (Hobson et al., 2000, p. 835) to depict the functional properties of what he calls "the dreaming brain" (Figure 10.2).

The problem with all of this is that the two disciplines have, until very recently, focused on such completely different aspects of the functional domains they share—and as a result they have had almost nothing to say to each other. But that has changed now. Neuroscientists are at last turning their attention to the *inner* workings of the mind. The chapters of this book have

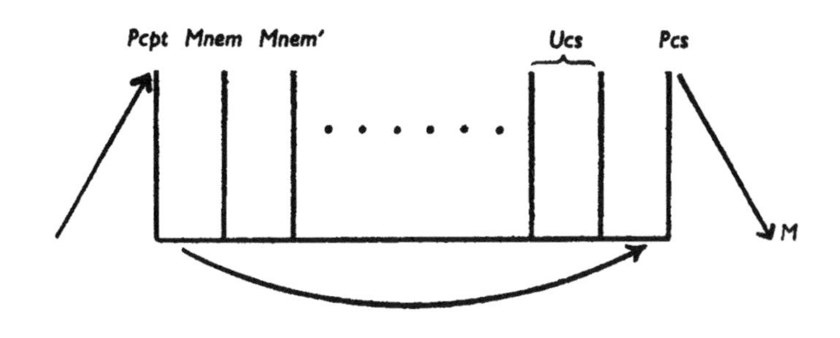

FIGURE 10.1
Functional depiction of the dreaming mind (from Freud, 1900a, p. 541)

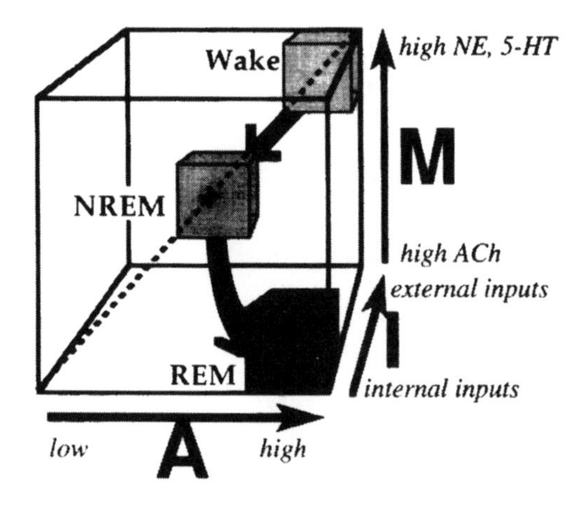

FIGURE 10.2
Functional depiction of the dreaming brain (from Hobson et al., 2000, p. 835)

presented—in brief outline—a survey of the resultant knowledge: a body of knowledge that truly deserves to be called the "neuroscience of subjective experience."

So now we are faced with *two* descriptions of the metapsychology of consciousness (and emotion, memory, dreams, and so on)—two descriptions of everything that makes up the chapter headings of this book. This is neither a happy nor a healthy situation. Clearly, there is only one thing to be done about it: *the two perspectives have to be integrated and combined, and thereby reconciled with one another.*

THE ADVANTAGES OF "OBJECTIVE" SCIENCE

There is no denying the fact that although psychoanalysis and neuroscience are both studying the same thing (from different perspectives), psychoanalytic knowledge is far less secure than that of neuroscience. Most educated people seem happy to accept that the latest theories emanating from cognitive neuroscience about the organization of memory (or consciousness, or emotion, etc.) are *scientific* theories—the most reliable knowledge that we currently have about the laws governing the aspect of the mental apparatus in question. This cannot be said for psychoanalytic theories. The discrepancy is obviously not attributable to some property of the mental apparatus itself—this is simply a piece of nature, and, as we have said already, *both* disciplines are studying this thing. The source of the discrepancy lies elsewhere—namely, in the *observational perspectives* that the two disciplines adopt. It is the observational perspective of psychoanalysis that makes its theories so unreliable.

The observational perspective of psychoanalysis generates data of a fleeting and fugitive kind. Subjective experience—the observational "stuff" of psychoanalysis—often seems impossible to pin down. This is inherent in the very nature of the stuff. What cannot be pinned down cannot be measured. That, too, is in the nature of the stuff. But, above all, subjective experience is *subjective*. That is to say, it is singular and unique, and it is only observable by the subject himself or herself. For this reason, no two observers of a subjective experience can ever be expected to agree; they are literally *unable* to observe the same thing. It is true that they can observe the same *type* of thing and can, thereby, at least compare their *conclusions* in order to reach consensual *generalizations*. However, for precisely this reason, the resultant knowledge (at least for topics of real interest) becomes increasingly insecure. The most generalizable conclusions are, by and large, the farthest removed from the actual object of study (the raw data of subjectivity).

Psychoanalysts may console themselves with the fact that the unreliability of their knowledge is a product of the way that the mind works, but this is not much of a consolation. For there is another branch of science that studies the same thing—"the mind"—and the conclusions of that other science seem so much more reliable.

This challenge to psychoanalysis can be turned to its advantage. Since the generalizations of psychoanalysis refer to the same thing as the generalizations of cognitive neuroscience, the one set of generalizations can be tested against the other. And this second set of data is "objective"—it concerns a physical "thing." It *can* therefore be pinned down, and measured, and scrutinized by as many independent observers as one wants.

THE ADVANTAGES OF "SUBJECTIVE" SCIENCE

It would be a grave mistake to conclude (as many are tempted to do) that mental science can therefore do without psychoanalysis. If psychoanalysis and cognitive neuroscience are studying the same thing, and the conclusions of cognitive neuroscience are so much more reliable, then who needs psychoanalysis?

The answer is that psychoanalysis gives us access to inner workings of the mental apparatus that cannot be studied—literally cannot be *seen*—from the "objective" point of view. Feelings are a perfect example. Feelings cannot be seen, but they most certainly *exist*. They are part of nature. And as such they exert effects on the *other* parts of nature, including those parts that can readily be seen.[1] Hence all the agony of the mind–body problem (see chapter 2). How can something immaterial affect something material, unless it is *real*? The answer is obvious: of course it *is* real! Reality is not synonymous with visibility. Feelings are real. They exist. They have effects. And for that reason, science ignores them at its peril.

[1] Think of suicide and murder as examples: feelings have effects.

A science that sought to understand the piece of nature that is the human being would be led seriously astray if it did not take account of the feelings (and phantasies and reminiscences and the like) that shape our inner lives: the choices we make, the things we do, the way we behave, *who we are*. The inner world of subjective experience, *as we experience it*, is as real as are apples and tables.

The claim of psychoanalysis to scientific interest is simply that. For all its faults, it makes a serious attempt to come to grips with this aspect of nature. It is an enrichment of science. It is an acknowledgment of reality. The complexities and difficulties of the inner world of subjective experience are part and parcel of the mind and how it works. For this reason, modern neuroscience has as much to gain from psychoanalysis as modern psychoanalysis has to gain from neuroscience.

PREJUDICE AGAINST NEUROSCIENCE

It would be a unfair to leave readers with the impression that neuroscientists are prejudiced against psychoanalysis (in fact, this is increasingly less the case; see below), whereas psychoanalysts, by contrast, are fearlessly open-minded and truth-loving. Sadly, this is not the case. Psychoanalysts are no less prejudiced against neuroscience than neuroscientists are against psychoanalysis.

In this, they blindly follow Freud. And in doing so, they misunderstand him! One hundred years ago, Freud (1900a, p. 536) famously asserted that he was "disregard[ing] the fact that the apparatus with which we are here concerned is also known to us in the form of an anatomical preparation," and implored his readers to "remain upon psychological ground." But that was one hundred years ago. Freud did not say that the apparatus he was describing was *not* also an anatomical preparation; he said only that he was *disregarding* that fact (Freud, 1915e, p. 174).[2] Moreover, he called only for a *period* of disciplinary independence, so that psychology may, "*for the present* . . . proceed according to its own requirements." Immediately he went on to add: "after we have completed our psychoanalytic work we shall have to find a point of contact with biology" (p. 175). He went further still:

> Biology is truly a land of unlimited possibilities. We may expect it to give us the most surprising information and we cannot guess what answers it will return in a few dozen years. . . . They may be of a kind which will blow away the whole of our artificial structure of hypotheses. [Freud, 1920g, p. 60]

That was written in 1920. It is now "a few dozen years" later. Freud's recommendation to his followers that they remain aloof of developments in neuroscience has clearly lapsed. It was only a *temporary strategy*, designed to allow the subjective perspective on the mind maximum scope for unexpected insights and discoveries, whereafter those discoveries can *and must* be reconciled with neurobiology.

Freud's followers appear to have forgotten that the isolation of psychoanalysis was a temporary strategy. They turned it into an article of faith, and psychoanalysis has suffered as a result. A century on, psychoanalysis has had an adequate opportunity to see how far it can get on its own, and the recent consensus—certainly among the greater community of scientists—is that the clinical method of psychoanalysis has taken us about as far as it is going to, on its own. It is time to "find a point of contact with biology."

[2] For a detailed exegesis of Freud's position on this whole question, see Solms and Saling (1986, 1990), Kaplan-Solms and Solms (2000), and Solms (2000b).

WHERE DO WE GO FROM HERE?

For the historical reasons just outlined, psychoanalysts today are ill-equipped to re-join with neuroscience. They simply do not have the knowledge and the skills; they have not kept abreast of the relevant scientific developments. Now it might appear to be too late. Of course it is daunting, when you have been a psychotherapist all your life, suddenly to recognize that neuroscientists know some important things about the inner workings of the mind that might impact directly on everything you do. Nobody wants to start all over again.

Similar considerations apply in the other direction too. The pressures to abandon psychoanalysis are substantial. Both of us [MS, OT] have been in professional situations—more often than we like to remember—where our interest in psychoanalysis has made it difficult to maintain the respect of our colleagues, the esteem of our students, and the willingness of journal editors to publish our work. (See, for example, Turnbull, 2000.) *We* think of psychoanalysis as offering a solution to a full half of the complex puzzle of the mind, and for this reason we are determined to use it. But in a climate of mutual hostility, that is not always easy. And the hostility really is *mutual*. Being prepared to revise or replace the cherished assumptions of psychoanalytic theory that fail to stand up to appropriate scientific scrutiny does not endear you to your psychoanalytic colleagues either! This, then, is the difficult position that we have to occupy—defending the value of a marginalized discipline, while simultaneously inviting it to enter the lion's den.

It would be wrong to overlook such prosaic reasons for the continued isolation of psychoanalysis. Such factors need to be acknowledged and confronted head-on. Psychoanalytically educated readers of this book have already taken the first step. They have overcome the first resistance that separates our two disciplines and have begun to acquaint themselves with the sort of knowledge that is needed to make interdisciplinary communication between the two fields possible. This is a very important step. It qualifies readers of this book to join the "advance guard" of like-minded people, emanating from the other initiatives that are currently being undertaken—which we shall now describe—to break down the old barriers and to construct a radically new approach to mental science.

FOUNDING A NEW DISCIPLINE

In the early 1990s, recognizing the situation just described, we joined with some like-minded colleagues who had established a small interdisciplinary study group under the auspices of the New York Psychoanalytic Institute. This group was convened by Drs. Arnold Pfeffer and James Schwartz, both leading figures in their respective fields, who happened to be friends and neighbors. The founding goal of their group was similar to the purpose of this book: cross-disciplinary education on topics of mutual interest.

The group held (and continues to hold) monthly meetings, each devoted to a different topic. At these meetings, a neuroscientist outlines the findings of his or her discipline on the topic at issue. Then a psychoanalytic perspective on the topic is presented. This highlights both the psychological complexities raised by the neuroscientific perspective and some of the problems that remain. In this way, a dialogue is begun.

The "Neuroscience Study Group" of the New York Psychoanalytic Institute (now called the Arnold Pfeffer Center for Neuro-Psychoanalysis) gradually evolved into a multifaceted center for research, training, and education at the interface of psychoanalysis and neuroscience. The founders of this center soon recognized that if they were going to make any substantial headway toward their goal of breaking down the barriers dividing their disciplines, they needed to enlist the public support of leading personalities in both fields. This, they felt, would confront directly the real obstacles, just outlined, that made it so difficult for junior members of either discipline to openly declare their interest in the "other side" and therefore devote the time and resources necessary to acquaint themselves properly with its methods, findings, and theories.

They decided that their study group would invite 20 prominent members of each field to form two Scientific Advisory Boards, which could simultaneously act as Editorial Advisory Boards to a new interdisciplinary journal that they wanted to establish. To our amazement and delight, 17 of the 20 neuroscientists that we invited, and 19 of the 20 psychoanalysts, accepted the invitation. In fact, we ended up in the embarrassing position of having almost too many advisors. We had expected that just a handful would accept and had planned to use them as bait to entice a few more![3]

[3]The current members of these boards are the following. *Neuroscience*: Eduardo Boncinelli, Joan Borod, Allen Braun, Jason Brown, Antonio Damasio, John DeLuca, Wolf-Dieter Heiss, Nicholas Humphrey, Eric Kandel, Marcel Kinsbourne, Joseph LeDoux, Benjamin Libet, Detlef Linke, Rudolfo Llinas, John C. Marshall, Jaak Panskepp, Michael Posner, Karl Pribram, V. S. Ramachandran, Oliver Sacks, Todd C. Sacktor, Michael Saling, Daniel Schacter, James Schwartz, Carlo Semenza, Tim Shallice, Wolf Singer, Max Velmans. *Psychoanalysis*: Jacob Arlow, Charles Brenner, Luis Chiozza, Peter Fonagy, Manuel Furer, Robert Galatzer-Levy, André Green, Ilse Grubrich-Simitis, Ernest Kafka, Otto Kernberg, Marianne Leuzinger-Bohleber, Fred Levin, David Milrod, Arnold Modell, David Olds, Barry Opatow, Mortimer Ostow, Morton Reiser, Allan Schore, Theodore Shapiro, Howard Shevrin, Albert Solnit, Riccardo Steiner, Arthur Valenstein, Daniel Widlöcher, Clifford Yorke.

WHAT HAS CHANGED?

Leading neuroscientists had become much more open to psychoanalysis than we expected, and psychoanalytic authorities were similarly far less cautious about neuroscience than their previous behavior had led us to expect. Clearly, something had changed. The main factors were quite obvious and, in retrospect, seem predictable. First, psychoanalysis had reached its limits. Freud's method had proved every bit as fruitful as he had hoped it would be for opening new perspectives on the mind and framing new hypotheses about its workings. But it had also proved hopelessly inadequate for the purposes of deciding between *competing* possibilities. The result was a gradual proliferation of incompatible theories, regarding seemingly unanswerable questions, each of which was defended by a rival theoretical "school." This situation, not surprisingly, led to plummeting public confidence. Thereafter, other forces intervened, and it soon became a stark matter of "adapt or die."

Simultaneously, technological advances in the neurosciences were opening new vistas. Advances in brain imaging (as well as molecular neurobiology) suddenly rendered anything—literally any aspect of mental life—fair game for experimental neuroscientific investigation. In psychology in general, the demise of behaviorism was also a major factor, coupled with the logical extension of the *cognitive* science revolution into the motivational and emotional domains that are so obviously inseparable from cognition. On both sides, therefore, albeit for very different reasons, there was a new openness to previously closed subjects. This was accompanied by a new humility and respect (again for different reasons) for the achievements of the "other side." In the case of psychoanalysis, as we have said, the humility seemed to be born of declining self-confidence, and the respect from an awareness of the contrasting fortunes of neuroscience—with the achievements of neuroscience having captured the public imagination in direct proportion to the demise of psychoanalysis. In the case of neuroscience, on the other hand, a new humility and respect for psychoanalysis was apparently born from the dawning realization of how difficult it is to study human subjectivity. The enormous methodological and conceptual problems that had haunted psychoanalysis from its inception had suddenly become neuroscientific problems too.

A JOURNAL, A CONGRESS, A SOCIETY, AN INSTITUTE

The success of the interdisciplinary study group in New York led to the rapid formation of numerous similar groups in other cities around the world.[4] As a consequence, the need arose for international contact, for the sharing of resources, findings, and experience.

This led initially to the establishment of the journal mentioned above, which adopted the format of the dialogues hosted by the New York Psychoanalytic Institute—that is, it published target articles on topics of mutual interest together with interdisciplinary peer commentaries. It was decided that the journal would be called *Neuro-Psychoanalysis*—and this, by default, seems to have become the name of the new discipline.[5]

The creation of the journal led to the idea of holding an international congress, now held annually. Each congress is devoted to a single interdisciplinary topic. The first congress, held in July 2000, on the topic of emotion, was opened by Oliver Sacks at the Royal College of Surgeons in London. The congress

[4] At the time of writing such groups exist in Ann Arbor, Bologna, Boston, Buenos Aires, Chicago, Cologne, Frankfurt am Main, Ghent, Jerusalem, London, New Haven, New York, Porto, Sao Paulo, Southern Brazil, Stockholm, Toronto, Vienna, and Washington.

[5] This journal is currently edited by Edward Nersessian and Mark Solms. From 2003 onward, it will be published by Karnac, an imprint of Other Press. For further details visit the journal's website at www.neuro-psa.com.

was spread over three days, with plenary presentations by Antonio Damasio, Jaak Panksepp, and Mark Solms. Attendance at this meeting was heavily oversubscribed. The second congress, on the topic of memory, was held in April 2001, at the much larger venue of the New York Academy of Medicine, and it was opened by Karl Pribram. The plenary speakers were Daniel Schacter, Elizabeth Loftus, and Mark Solms. The third congress is to be held in September 2002 in Stockholm, on the scheduled topic of sexuality and gender.[6]

At the London congress an international society was formed, which is now called the International Neuro-Psychoanalysis Society. There were over 400 founding members. We are currently in the process of establishing a parallel Neuro-Psychoanalysis Institute, which will take financial and administrative responsibility for the research, publishing, and educational needs of the fledgling discipline. A great deal has been achieved in a very short period of time, and the future looks increasingly bright.

PSYCHOANALYSIS WELCOMED BACK INTO THE FAMILY OF SCIENCES

One measure of the new state of affairs can be judged from two recent articles by Eric Kandel (1998, 1999).[7] He wrote these articles, by invitation, for the prestigious *American Journal of Psychiatry*. In these remarkable papers he noted that "psychoanalysis still represents the most coherent and intellectually satisfying view of the mind that we have" (1999, p. 505). He

[6] For further information on these congresses and on the Society described below, contact paula.barkay@annafreud.org.

[7] Kandel is one of the world's foremost neuroscientists, author—with James Schwartz, mentioned above—of one of its most respected textbooks (Kandel, Schwartz, & Jessell, 2000), and the most recent winner of the Nobel Prize in Medicine and Physiology. Some of Kandel's work is described in chapter 4.

argued, on this basis, that psychoanalysis provides neuroscientists with the most appropriate theoretical starting point to tackle the complexities and difficulties just mentioned. In essence, he called for a merging of psychoanalysis and neuroscience, and he saw this merger as a "new intellectual framework for psychiatry" in the twenty-first century.

The battle is by no means won, however, and psychoanalysis still faces many of its old resistances and prejudices. Not all of these, in truth, are unjustified. In the midst of the praise and affection for psychoanalysis cited above, Kandel also contended that the once innovative tool of its clinical method has "exhausted much of its novel investigative power" (1999, p. 506) and needs to be bolstered by nonclinical techniques. He also urged psychoanalysts to move away from single-case studies toward "more reliable" methods of observation and "greater experimental control," because "although psychoanalysis has historically been scientific in its aim, it has rarely been scientific in its methods" (p. 506). He especially stressed the role of "blind" data collection, and he praised recent work by psychoanalytic researchers who have reported behavioral observations instead of "hearsay evidence" (p. 506).

However, the uncertainty and doubt that still plague psychoanalysis are nothing to be ashamed of. Psychoanalysis has come a great distance in a hundred years, which is not a very long time in science. It is therefore not surprising that there still are question marks attached to just about everything that it has revealed. Linking the metapsychological conclusions of psychoanalysis with the equivalent conclusions of neuroscience—derived, as they are, from another observational perspective on the same thing, and an "objective" one at that—provides psychoanalysis with unlimited opportunities for overcoming its uncertainties and doubts. In doing so it has nothing to lose, for the integration of psychoanalytic knowledge with the neuroscientific equivalent is by no means tantamount to a replacement or reduction of psychoanalytic knowledge to neuroscience. Nobody gains anything by jettisoning the subjective perspective of psychoanalysis; our goal is only to strengthen it, by coupling it to another, parallel perspective, which has a different set of weaknesses, so that the two perspectives may serve as mutual correctives for viewpoint-dependent errors.

WHAT CAN WE LEARN FROM A DIALOGUE?

Dialogues of the kind described above have their limits too. They are useful for educating the participants about each other's fields, and for correcting misconceptions. But dialogues are not research, and speculation is not science. Dialogues cannot make new discoveries.

The problem is essentially one of methodology. A dialogue begins and ends with two different points of view, notwithstanding any points of contact. Descriptions of abstract entities starting from different observational perspectives are inevitably couched in different terms and concepts, which will not map onto each other in any simple way. There is an enormous temptation to fudge this issue. When analysts talk about "drives" or "arousal" or "inhibition," they seldom mean the same thing as neurobiologists using the same words. Freud's early attempt to neurologize the mind in the abstract—in his celebrated "Project for a Scientific Psychology" (1950 [1895])—was doomed from the start: it truly was an "aberration," as Freud himself concluded (1950a [1887–1902], p. 134).[8] Scientific models cannot be integrated in *theory*—one needs to determine (by laborious scientific *observation* and *experimentation*) where and whether they refer

[8]Freud described his methodology in the 1895 "Project" as one of "imaginings, transpositions and guesses"—and that is exactly the problem.

to the *same things* and then codify the dual referents into a new, integrated language.

What is needed is a *method* by means of which one and the same thing can be studied simultaneously from both the psychoanalytic and the neuroscientific perspectives, so that one can be sure that the two sets of observations (and the resultant theoretical accounts) refer to the same piece of reality. Only this enables us to link the two theories in *reality* rather than in words.

A RECOMMENDED METHOD

Such a method exists. In fact we have already made use of its findings in the pages of this book. In chapter 8, for example, we considered whether the functions of the system *Ucs* and those of the right cerebral hemisphere were, as some authors have speculatively claimed, synonymous. We decided this question by studying *actual cases* of right-hemisphere damage and by observing psychoanalytically whether or not the predicted effects materialized. They did not. We therefore concluded that the two abstractions (functions of the *Ucs* and functions of the right hemisphere) are not, in fact, synonymous. It is as simple as that.

This is the well-established clinico-anatomical method, by now very familiar to our readers. People who suffer brain tumors, strokes, and so forth are *people* just like ourselves—they have developed personalities, complex histories, and rich internal worlds. Since these things are the stuff of psychoanalysis, these people can be studied psychoanalytically just like anyone else. In this way, basic clinico-anatomical correlations can be drawn, directly linking psychoanalytic variables with neurological ones and thereby integrating them with each other on a valid empirical (rather than theoretical) basis.

Our approach is simply to take these patients (whose mental changes behavioral neuroscientists traditionally studied using pencil-and-paper tests) into psychotherapeutic treatment. This enables us to operationalize, in an ordinary psychoanalytic way, the psychological variables that interest us—while we simultaneously do whatever we can to help them come to terms with what has happened to them (for detailed case examples see Kaplan-Solms & Solms, 2000). In doing so, we are not doing anything different from what psychotherapists always do. This approach to brain–mind correlation does not, therefore, require anyone to "start all over again" in doing research in this field.

MULTIPLE OBSERVATIONS

On the basis of this sort of psychoanalytic investigation, one can determine whether, and in what way, a particular function of the mental apparatus has been affected by a brain lesion—for example, the function of "secondary-process" inhibition. We can then correlate the observed changes with the part of the brain that was damaged. This reveals the contribution that the part of the brain in question made to the organization of that mental function. If, for example, we observe that patients with ventromesial frontal-lobe damage suffer a near-total breakdown of secondary-process inhibition, we may reasonably conclude that this psychoanalytic function is coextensive with the neuropsychological functions of the ventromesial frontal region.

This approach assumes that the correlation between the observed lesion and the observed mental change was not simply a coincidence. That assumption is tested by checking one's observations in the individual case against analogous observations in many other cases (the more the better) with damage to the same part of the brain. In this respect, neuro-psychoanalytic research is no different from any other branch of neuropsychological research. By investigating *groups* of patients, it is possible to discern reliable patterns of association between brain regions and mental functions of psychoanalytic interest. Kaplan-Solms and Solms (2000) describe three small groups of this sort, for three separate brain regions, and the results appear to be quite reliable.

Progress has, however, been slow, given the nature of the method, and this is why we are keen to encourage other psychotherapists to participate in this important research endeavor. Once again, note that what is particularly valuable about this method is that the psychotherapist does not need to do anything fundamentally different from usual. This is not a technologically driven method. It requires no laboratory, nor advanced knowledge about the brain. All that one needs is an open analytical mind. It may well be that psychotherapists could benefit from the diagnostic assistance of a neuropsychologist, to interpret the more neurological aspects of the case, but the bulk of the problem can be dealt with in the usual way.

OTHER SUITABLE METHODS

Having recommended a basic method for neuro-psychoanalysis, we hasten to add that it is not the *only* way of making progress in this field. Many other neuroscientific methods, quite distinct from the clinico-anatomical method, also embody the requisite principles. The most obvious example would be to study the mental effects of psychopharmacological agents (psychiatric drugs) that alter brain chemistry. Studying patients who use such drugs permits, once again, correlative observations between neural variables (in this instance, neurochemical ones) and psychoanalytic variables. Across a range of patients, systematically investigated, this should enable one to establish empirical links between, say, decreased internal aggression and decreased serotonin reuptake (Zueler & Maas, 1994). Serotonin is something that psychiatrists are manipulating all the time, and it is nothing more than a matter of good experimental design to investigate systematically the effects of such manipulations on variables of interest to psychoanalysts.[9]

There has been a desire (especially in America, where the most progress has taken place) to use the most technologically sophisticated methods available to address these problems. People want to use the latest techniques and to see technicolor images. They want to elucidate the neural organization of particular mental mechanisms (e.g., "repression"), or whole psychopathologies (e.g., "hysteria"), or even whole functional systems (e.g., "the Unconscious") by trying to visualize them in PET or fMRI scanners. In fact, it is not possible to do that with any degree of accuracy. This is not because such things cannot be visualized (the neural correlates of literally any mental entity can be visualized), but because the mental entities in question cannot be artificially operationalized and manipulated in the laboratory conditions that such imaging techniques require. In the past, all sorts of laboratory methods have been used to investigate psychoanalytic concepts (for a review see Fisher & Greenberg, 1995), but they have never properly operationalized the phenomena with which psychoanalytic theory deals.

Psychoanalytic concepts are operationalized in the *clinical* situation. Using expensive scanners now will not avoid the errors of the past, as the financial cost of a technology does not protect one from the flaws of experimental design. A salutary reminder is the fact that there are, at present, still great controversies about a number of findings of functional-imaging studies even in mainstream neuropsychology. Here, the anatomical correlates of a wide range of psychological functions are very well known, and

[9] Mortimer Ostow has undertaken a lifetime of pioneering neuro-psychoanalytic work using this method. See Ostow (1962).

yet disagreement still remains (for a critical review of the field see Bub, 2000).[10]

Nevertheless, imaging could very well prove highly useful once we have gained a firmer foothold on the basic problems using the older techniques. Achieving an empirically based understanding of how the two fields relate to each other in broad-brush-stroke clinical terms paves the way for more precise studies of more circumscribed questions. Perhaps a brief example (building on matters discussed in chapter 6) is appropriate here.

The neuropsychology of dreams: From the general to the specific

We discovered, using the clinico-anatomical method (Solms, 1997a), that one brain region that is necessary for the generation of dreams is the white matter of the ventromesial quadrant of the frontal lobes (Figure 3.4). Damage to this region leads to a cessation of dreaming. From the perspective of clinical observation, we observed that these nondreaming patients were also

[10] Perhaps the best-known example is that of the hippocampus. We have known for decades, from literally thousands of studies using a range of techniques, that the hippocampus is absolutely central to memory formation (see chapter 5). It is a finding well beyond dispute. However, for many years functional-imaging studies suggested that the hippocampus was *not* active during memory tasks (compared with baseline, nonmemory, conditions). If we had relied on functional-imaging findings alone, this finding would have led a fledgling discipline (like neuro-psychoanalysis) up a dead-end street. It was only the rock-solid knowledge of conventional neuropsychology (clinico-anatomical correlation) that allowed memory researchers to tolerate this bizarre finding for a number of years, until the aberrant results were finally explicable. In fact, it now seems likely that the hippocampus is almost *always* active, whether you are consciously aware of being engaged in a memory task, or not, and the imaging findings are once again congruent with those in the rest of memory research (for more information see Martin, 1999).

aspontaneous, inert, and apathetic. Studied psychoanalytically, they showed a massive depletion in what used to be called libido (appetitive interest). Establishing this sort of broad relationship made it possible for us to find our bearings as to how the "libido" concept might *begin* to map onto brain anatomy and chemistry. We were now in a position to ask more precise questions—not about which "quadrant" of the brain is the vehicle of libidinal interest, but which specific fiber pathway it courses through. This is being done using fMRI imaging in normal subjects, and specific pharmacological probes. Preliminary results suggest that it is the mesocortical–mesolimbic dopamine system in particular, connecting the ventral tegmental area with the nucleus accumbens, that is the crucial component of the ventromesial-quadrant lesion site. That conclusion can now be checked, in turn, using a host of other high-technology methods. In this way, we are isolating the neural correlates of an important component of the "libido" concept.

The basic clinico-anatomical method therefore does not exclude the use of more technologically sophisticated methods. Each method has its proper place. At present, we are very much at the beginning of trying to achieve a broad understanding of the relationship between our two different models of the mind. Much of this scientific investigation can proceed as part of what is ordinary clinical work—which matches the way that neuropsychology established itself some one hundred years ago. As we proceed, specialists will want to take some questions further, but we should not try to run before we can walk.

"TESTING" PSYCHOANALYTIC THEORIES

One final point. There is naturally a desire to use (or to use too quickly) neuroscientific methods to "test" psychoanalytic theories: "Is there really such a thing as repression?" "Are dreams really motivated by wishes?" Such tests may well be possible in the future, and one of the many promising prospects of neuro-psychoanalysis is that it *should* allow us to ask and answer such metapsychological questions—questions that cannot be satisfactorily resolved by the psychoanalytic method alone.

But an intermediate step needs to be undertaken before such questions can even be asked. Although some people may regard this phase as less interesting, and perhaps even reactionary, it is unquestionably a necessary step. We first have to establish the neurological *correlates* of the metapsychological concepts that make up psychoanalytic theories, before we can begin to *test* the theories themselves. We have to see where the pieces of the theory lie in the tissues and processes of the brain before we can systematically investigate them and experiment on them in the controlled ways required for systematic hypothesis testing. For example, if we want to test the wish-fulfillment theory of dreams, we first have to find the neural correlates of the various component parts of that theory. Anything resembling a complete list of these will have to include libidinal drive, censorship, regression, reality testing, the perceptual systems, and more. Once we have some broad idea of how these concepts correlate with brain anatomy and chemistry, we can then begin to take the second step of asking whether they relate to each other in physical reality in the way that psychological theory predicts. Without the earlier, correlative phase, we run the risk of testing apples by measuring pears.

CLOSING REMARKS

This concludes our introduction to the neuroscience of subjective experience. We stand at the dawn of an exciting new era in mental science. All sorts of possibilities are opening up. We appear, at last, to have within our grasp the possibility of studying in measurable, physical units, the inner life of the mind—the traditional preserve of psychoanalysis.

Despite a century of concerted effort, psychoanalysts have not been able to convince the scientific community at large that they have truly revealed the laws that govern this most wonderful and mysterious part of nature: our very own selves. The core of this book (chapters 3 to 9) was designed to demonstrate that a substantial body of neurobiological knowledge now exists that concerns many issues of traditional interest to psychoanalysts. This situation represents something of a crossroads for psychoanalysis. Psychoanalysts can choose to remain aloof from neuroscience for another hundred years, but we have little doubt that that would be to the detriment of both psychoanalysis and neuroscience. There is only one mental apparatus. In the long term, a comprehensive neuroscience of subjective experience will be developed, with or without psychoanalysis. The cooperation of psychoanalysts at this point will surely speed up the process and enrich it immeasurably, but science has a way of eventually finding a route through the darkest forests, and it will no doubt do the same with this one in the end.

The high road for psychoanalysis is to engage with the neuroscientific issues that should now directly interest it. This will not be an easy task. Most psychoanalysts are unfamiliar with the complexities of neuroscience, and (one must admit) they are often poorly equipped to design and implement systematic scientific investigations. Some psychoanalysts today are, however, keen to rise to the challenge, and this book is designed to aid those who wish to do so. If a critical mass of psychoanalysts should choose this path, there is much to be gained in return for the effort that it will involve. A radically different psychoanalysis will emerge. It will be a psychoanalysis that retains its pride of place as the science of human *subjectivity*—the discipline through which we investigate the stuff of individual experience: the living of a life. But its claims will be far more securely grounded. We will better understand how mental disorders arise. We will be able to target our therapies at those who can benefit most, and in the ways that work best. We may extend our clinical reaches in previously undreamt-of directions. And in the end, we believe, we shall be able to say with confidence at last: this is how the mind *really* works.

REFERENCES

Adolphs, R., Tranel, D., and Damasio, A. R. (1994). Impaired recognition of emotion in facial expressions following bilateral damage to the human amygdala. *Nature, 269*: 669–672.

Anderson, M. C., and Green, C. (2001). Suppressing unwanted memories by executive control. *Nature, 410*: 366–369.

Anderson, S. W., Bechara, A., Damasio, H., Tranel, D., and Damasio, A. (1999). Impairment of social and moral behaviour related to early damage in human prefrontal cortex. *Nature Neuroscience, 2*: 1032–1037.

Aserinsky, E., and Kleitman, N. (1953). Regularly occurring periods of eye motility and concomitant phenomena during sleep. *Science, 118*: 273–274.

Baars, B. J., and McGovern, K. A. (1999). Consciousness cannot be limited to sensory qualities: Some empirical counterexamples. *Neuro-Psychoanalysis, 2*: 11–13.

Baddeley, A. (1997). *Human Memory: Theory and Practice.* London: Psychology Press.

Bailey, J. M., Willerman, L., and Parks, C. (1991). A test of the maternal stress theory of human male homosexuality. *Archives of Sexual Behaviour, 20*: 277–293.

Bakker, A., Van Balkom, A. J., and Van Dyck, R. (2001). Comparing psychotherapy and pharmacotherapy. *American Journal of Psychiatry, 158*: 1164–1166.

Bargh, J. A., and Chartrand, T. L. (1999). The unbearable automaticity of being. *American Psychologist, 54*: 462–479.

Baxter, L. R., Jr., Schwartz, J. M., Bergman, K. S., Szuba, M. P., Guze, B. H., Mazziotta, J. C., Alazraki, A., Selin, C. E., Ferng, H. K., Munford, P., et al. (1992). Caudate glucose metabolic rate changes with both drug and behavior therapy for obsessive-compulsive disorder. *Archives of General Psychiatry, 49*: 681–689.

Bechara, A., Damasio, A. R., Damasio, H., and Anderson, S. W. (1994). Insensitivity to future consequences following damage to human prefrontal cortex. *Cognition, 50*: 7–15.

Bechara, A., Damasio, H., and Damasio, A. R. (2000). Emotion, decision making and the orbitofrontal cortex. *Cerebral Cortex, 10*: 295–307.

Bradshaw, J. L., and Mattingley, J. B. (1995). *Clinical Neuropsychology: Behavioural and Brain Science.* San Diego, CA: Academic Press.

Braun, A. (1999). The new neuropsychology of sleep. *Neuro-Psychoanalysis, 1*: 196–201.

Braun, A., Balkin, T., Wesenten, N., Carson R., et al. (1997). Regional cerebral blood flow throughout the sleep–wake cycle. *Brain, 120*: 1173–1197.

Braun, A., Balkin, T., Wesenten, N., Gwadry, F., et al. (1998). Dissociated pattern of activity in visual cortices and their projections during human rapid eye movement sleep. *Science, 279*: 91–95.

Brody, A. L., Saxena, S., Schwartz, J. M., Stoessel, P. W., Maidment, K., Phelps, M. E., and Baxter, L. R. Jr. (1998). FDG-PET predictors of response to behavioral therapy and pharmacotherapy in obsessive compulsive disorder. *Psychiatry Research, 84* (November, No. 1): 1–6.

Bub, D. N. (2000). Methodological issues confronting PET and fMRI studies of cognitive function. *Cognitive Neuropsychology, 17*: 467–484.

Carey, D. P. (1996). "Monkey see, monkey do" cells. *Current Biology, 6*: 1087–1088.

Chalmers, D. (1995). *The Conscious Mind: In Search of a Fundamental Theory.* New York: Oxford University Press.

Claparède, E. (1911). Reconnaissance et moitié. *Archives de Psychologie, 11*: 79–90.

Crick, F. (1994). *The Astonishing Hypothesis: The Scientific Search for the Soul.* New York: Simon & Schuster.

Crick, F., and Koch, K. (2000). The unconscious homunculus. *Neuro-Psychoanalysis, 2*: 3–11.

Crick, F., and Mitchison, G. (1983). The function of dream sleep. *Nature, 304*: 111–114.

Damasio, A. (1994). *Descartes' Error.* New York: Grosset/Putnam.

Damasio, A. (1996). The somatic marker hypothesis and the possible functions of the prefrontal cortex. *Philosophical Transactions of the Royal Society of London (Biology), 351*: 1413–1420.

Damasio, A. (1999a). Commentary on Panksepp. *Neuro-Psychoanalysis, 1*: 38–39.

Damasio, A. (1999b). *The Feeling of What Happens.* London: Heinemann.

Damasio, H., Grabowski, T., Frank, R., Galaburda, A., and Damasio, A. (1994). The return of Phineas Gage: The skull of a famous patient yields clues about the brain. *Science, 264*: 1102–1105.

Dement, W., and Kleitman, N. (1957). Cyclic variations in EEG during sleep and their relation to eye movements, body motility, and dreaming. *Electroencephalography and Clinical Neurophysiology, 9*: 673–690.

DeRenzi, E. (1982). *Disorders of Space Exploration and Cognition.* Norwich: John Wiley.

Dorner, G., et al. (1980). Prenatal stress as a possible aetiogenic factor of homosexuality in human males. *Endokrinologie, 75*: 365–386.

Edelman, G. (1989). *The Remembered Present.* New York: Basic Books.

Engel, A., Kreiter, A., Koenig, P. M., and Singer, W. (1991). Interhemispheric synchronization of oscillatory neuronal responses in cat visual cortex. *Science, 252*: 1177–1179.

Feinberg, T. E., and Farah, M. J. (1997). *Behavioral Neurology and Neuropsychology.* New York: McGraw-Hill.

Ferng, H. K., Munford, P., et al. (1992). Caudate glucose metabolic rate changes with both drug and behavior therapy for obsessive-compulsive disorder. *Archives of General Psychiatry, 49*: 681–689.

Fisher, S., and Greenberg, R. P. (1995). *Freud Scientifically Reappraised: Testing the Theories and Therapy.* New York: Wiley-Interscience.

Foulkes, D. (1962). Dream reports from different stages of sleep. *Journal of Abnormal and Social Psychology, 65*: 14–25.

Freud, S. (1886). Über den Ursprung des Nervus acusticus. *Monatsschrift für Ohrenheilkunde, NF, 20* (8): 245; (9): 277.

Freud, S. (1891b). *On Aphasia.* London: Imago.

Freud. S. (1895d) with Breuer, J. *Studies on Hysteria. S.E., 2.*

Freud, S. (1900a). *The Interpretation of Dreams. S.E., 4 & 5.*

Freud, S. (1911b). Formulations on the two principles of mental functioning. *S.E., 12*: 215.

Freud, S. (1912–13). *Totem and Taboo. S.E., 13.*

Freud, S. (1913m [1911]). On psycho-analysis. *S.E., 12*: 207.

Freud, S. (1915c). Instincts and their vicissitudes. *S.E., 14*: 111.

Freud, S. (1915e). The unconscious. *S.E. 14*: 161.

Freud, S. (1917e [1915]). Mourning and melancholia. *S.E., 14*: 239.

Freud, S. (1920g). *Beyond the Pleasure Principle. S.E., 18*: 7.

Freud, S. (1923b). *The Ego and the Id. S.E., 19*: 3.

Freud, S. (1924b [1923]). Neurosis and psychosis. *S.E., 19*: 149.

Freud, S. (1930a). *Civilization and Its Discontents. S.E., 21*: 59

Freud, S. (1933a). *New Introductory Lectures on Psycho-Analysis. S.E., 22*: 3.

Freud, S. (1940a [1938]). *An Outline of Psycho-Analysis. S.E., 23*: 141.

Freud, S. (1950 [1895]). Project for a scientific psychology. *S.E., 1*: 175.

Freud, S. (1950a [1887–1902]). *The Origins of Psycho-Analysis.* New York: Basic Books.

Fridja, N. H., Manstead, A. S. R., and Bem, S. (2000). *Emotions and*

Beliefs: How Feelings Influence Thoughts. Cambridge: Cambridge University Press.

Galin, D. (1974). Implications for psychiatry of left and right cerebral specialisation: A neurophysiological context for unconscious processes. *Archives of General Psychiatry, 31*: 572–583.

Gallese, V., Fadiga, L., Fogassi, L., and Rizzolatti, G. (1996). Action recognition in the premotor cortex. *Brain, 119*: 593–609.

Gallwey, W. T. (1986). *The Inner Game of Golf.* London: Pan.

Gottesman, C. (1999). Neurophysiological support of consciousness during waking and sleep. *Progress in Neurobiology, 59*: 469–508.

Gray, C., Koenig, P., Engel, A., and Singer, W. (1989). Oscillatory responses in cat visual cortex exhibit inter-columnar synchronization which reflects global stimulus properties. *Nature, 338*: 334–337.

Gray, C., and Singer, W. (1989). Stimulus-specific neuronal oscillations in orientation columns of cat cortex. *Proceedings of the National Academy of Science USA, 86*: 1698–1702.

Hamer, D. H., Hu, S., and Magnuson, V. L. (1993). A linkage between DNA markers on the X-chromosome and male sexual orientation. *Science, 261*: 321–337.

Harlow, J. (1868). Recovery from passage of an iron bar through the head. *Massachusetts Medical Society Publications, 2*: 327–347.

Hartmann, E., Russ, D., Oldfield, M., Falke, R., and Skoff, B. (1980). Dream content: Effects of L-DOPA. *Sleep Research, 9*: 153.

Hebb, D. O. (1949). *Organization and Behaviour.* New York: Wiley.

Heilman, K., and van den Abell, T. (1980). Right hemisphere dominance for attention: the mechanisms underlying hemispheric asymmetries of attention (neglect). *Neurology, 30*: 327–330.

Heilman, K., and Valenstein, E. (1985). *Clinical Neuropsychology.* Oxford: Oxford University Press.

Hobson, J. A. (1988). *The Dreaming Brain.* New York: Basic Books.

Hobson, J. A. (1999). The new neuropsychology of sleep: Implications for psychoanalysis. *Neuro-Psychoanalsyis, 1*: 157–183.

Hobson, J. A., and McCarley, R. (1977). The brain as a dream state generator: An activation-synthesis hypothesis of the dream process. *American Journal of Psychiatry, 134*: 1335–1348.

Hobson, J. A., McCarley, R., and Wyzinki, P. (1975). Sleep cycle oscillation: Reciprocal discharge by two brainstem neuronal groups. *Science, 189*: 55–58.

Hobson, J. A., Pace-Schott, E. F., and Stickgold, R. (2000). Dreaming and the brain: Towards a cognitive neuroscience of conscious states. *Behavioral and Brain Sciences, 23*: 793–842.

Jones, B. (1979). Elimination of paradoxical sleep by lesions of the pontine gigantocellular tegmental field in the cat. *Neuroscience Letters, 13*: 285–293.

Jones, E. (1953). *Sigmund Freud: Life and Work, Vol. 1*. London: Hogarth Press.

Jouvet, M. (1967). Neurophysiology of the states of sleep. *Physiological Reviews, 47*: 117–177.

Kandel, E. R. (1998). A new intellectual framework for psychiatry. *American Journal of Psychiatry, 155*: 457–469.

Kandel, E. R. (1999). Biology and the future of psychoanalysis: A new intellectual framework for psychiatry revisited. *American Journal of Psychiatry, 156*: 505–524.

Kandel, E. R., Schwartz, J. H., and Jessell, T. M. (2000). *Principles of Neural Science*. Norwalk, CT: Appleton & Lange.

Kaplan-Solms, K., and Solms, M. (2000). *Clinical Studies in Neuro-Psychoanalysis*. London: Karnac Books.

Kolb, B., and Wishaw, I. P. (1990). *Fundamentals of Human Neuropsychology*. Washington, DC: W. H. Freeman.

Kondo, T., Antrobus, J., and Fein, G. (1989). Later REM activation and sleep mentation. *Sleep Research, 18*: 147.

Kosslyn, S. (1994). *Image and Brain*. Cambridge, MA: MIT Press.

LeDoux, J. (1996). *The Emotional Brain*. London: Weidenfeld & Nicolson.

LeVay, S. (1991). A difference in hypothalamic structure between heterosexual and homosexual men. *Science, 253*: 1034–1037.

LeVay, S. (1994). *The Sexual Brain*. Cambridge, MA: MIT Press.

Lickey, M. E., and Gordon, B. (1997). *Medicine and Mental Illness: The Use of Drugs in Psychiatry*. Washington, DC: W. H. Freeman.

Luria, A. R. (1966). *Higher Cortical Functions in Man*. New York: Basic Books.

Luria, A. R. (1973). *The Working Brain*. Harmondsworth: Penguin.

Luria, A. R., and Yudovich, F. (1971). *Speech and the Development of Mental Processes in the Child*. Harmondsworth: Penguin.

MacLean, P. (1949). Psychosomatic disease and the visceral brain: recent developments bearing on the Papez theory of emotion. *Psychosomatic Medicine, 11*: 338–353.

Martin, A. (1999). Automatic activation of the medial temporal lobe during encoding: Lateralized influences of meaning and novelty. *Hippocampus, 9*: 62–70.

McCarthy, R. A., and Warrington, E. K. (1990). *Cognitive Neuropsychology: A Clinical Introduction*. New York: Academic Press.

Mesulam, M.-M. (1981). A cortical network for directed attention and neglect. *Annals of Neurology, 10*: 309–325.

Mesulam, M.-M. (1998). From sensation to cognition. *Brain, 121*: 1013–1052.

Moss, A. D., and Turnbull, O. H. (1996). Hatred of the hemiparetic limbs (misoplegia) in a 10-year-old child. *Journal of Neurology, Neurosurgery and Psychiatry, 61*: 210–211.

Newman, J., and Baars, B. J. (1993). A neural attentional model for access to consciousness: A global workspace perspective. *Concepts in Neuroscience, 4* (2): 255–290.

Nielsen, T. A. (2000). A review of mentation in REM and NREM sleep: "Covert" REM sleep as a possible reconciliation of two opposing methods. *Behavioral and Brain Sciences, 23*: 851–866.

Ogden, J. A. (1996). *Fractured Minds: A Case-Study Approach to Clinical Neuropsychology.* New York: Oxford University Press.

Olds, J., and Milner, P. (1954). Positive reinforcement produced by electrical stimulation of septal area and other regions of rat brain. *Journal of Comparative and Physiological Psychology, 47*: 419–427.

Ostow, M. (1962). *Drugs in Psychoanalysis and Psychotherapy.* New York: Basic Books.

Pace-Schott, E., Solms, M., Blagrove, M., and Harnad, S. (Eds.) (in press). *Sleep and Dreaming: Scientific Advances and Reconsiderations.* New York: Cambridge University Press.

Panksepp, J. (1985). Mood changes. In: P. Vinken, G. Bruyn, and H. Klawans (Eds.), *Handbook of Clinical Neurology, 45* (pp. 271–285). Amsterdam: Elsevier.

Panksepp, J. (1998). *Affective Neuroscience: The Foundations of Human and Animal Emotions.* New York: Oxford University Press.

Raine, A., Lencz, T., Bihle, S., LaCasse, L., and Colletti, P. (2000). Reduced prefrontal grey matter volume and reduced autonomic activity in anti-social personality disorder. *Archives of General Psychiatry, 57*: 119–127.

Ramachandran, V. S. (1994). Phantom limbs, neglect syndromes, repressed memories, and Freudian Psychology. *International Review of Neurobiology, 37*: 291–333.

Ramachandran, V. S., and Blakslee, S. (1998). *Phantoms in the Brain: Human Nature and the Architecture of the Mind.* London: Fourth Estate.

Rizzolatti, G., and Arbib, M. A. (1998). Language within our grasp. *Trends in Neuroscience, 21*: 188–194.

Rizzolatti, G., Fadiga, L., Fogassi, L., and Gallese, V. (1999). Resonance behaviors and mirror neurons. *Archives of Italian Biology, 137*: 85–100.

Robertson, I. H., and Marshall, J. (1993). *Unilateral Neglect: Clinical and Experimental Studies.* Mahwah, NJ: Lawrence Erlbaum Associates.

Rogers, L. (1999). *Sexing the Brain.* New York: Columbia University Press.

Sacks, O. (1984). *A Leg to Stand On*. London: Duckworth.

Sacks, O. (1985). *The Man Who Mistook His Wife for a Hat*. London: Picador.

Schacter, D. (1996). *Searching for Memory*. New York: Basic Books.

Schore, A. (1994). *Affect Regulation and the Origin of the Self*. Mahwah, NJ: Lawrence Erlbaum Associates.

Schwartz, J. M., Stoessel, P. W., Baxter, L. R., Jr., Martin, K. M., and Phelps, M. E. (1996). Systematic changes in cerebral glucose metabolic rate after successful behavior modification treatment of obsessive-compulsive disorder. *Archives of General Psychiatry, 53*: 109–113.

Scoville, W. B., and Milner, B. (1957). Loss of recent memory after bilateral hippocampal lesions. *Journal of Neurology, Neurosurgery and Psychiatry, 20*: 11–21.

Searle, J. (1995a). The mystery of consciousness. *New York Review of Books, 42* (17): 60–66.

Searle, J. (1995b). The mystery of consciousness. *New York Review of Books, 42* (18): 54–61.

Schindler, R. (1953). Das Traumleben der Leukotomierten [The dream life of the leukotomized]. *Wiener Zeitschrift für die Nervenheilkunde, 6*: 330.

Snyder, S. H. (1999). *Drugs and the Brain*. New York: Scientific American Library.

Solms, M. (1995). New findings on the neurological organisation of dreaming: Implications for psychoanalysis. *Psychoanalytic Quarterly, 64*: 43–67.

Solms, M. (1997a). *The Neuropsychology of Dreams*. Mahwah, NJ: Lawrence Earlbaum Associates.

Solms, M. (1997b). What is consciousness? *Journal of the American Psychoanalytical Association, 45*: 681–778.

Solms, M. (2000a). Dreaming and REM sleep are controlled by different brain mechanisms. *Behavioral and Brain Sciences, 23*: 843–850.

Solms, M. (2000b). Freud, Luria and the clinical method. *Psychoanalysis and History, 2*: 76–109.

Solms, M. (2000c). A psychoanalytic perspective on confabulation. *Neuro-Psychoanalysis, 2*: 133–138.

Solms, M., and Saling, M. (1986). On psychoanalysis and neuroscience: Freud's attitude to the localisationist tradition. *International Journal of Psycho-Analysis, 67*: 397.

Solms, M., and Saling, M. (1990). *A Moment of Transition: Two Neuroscientific Articles by Sigmund Freud*. London: Karnac Books and The Institute of Psycho-Analysis.

Springer, S. P., and Deutsch, G. (1998). *Left Brain, Right Brain: Perspectives from Cognitive Neuroscience.* New York: W. H. Freeman.

Squire, L. R. (1987). *Memory and Brain.* New York: Oxford University Press.

Stein, B., and Meredith, M. A. (1993). *The Merging of the Senses.* Cambridge, MA: MIT Press.

Strawson, G. (1996). *Mental Reality.* Cambridge, MA: MIT Press.

Turnbull, O. H. (1996). Neuropsychological rehabilitation: Modern approaches and their debt to Aleksandr Luria. In S. Della Sala, C. Marchetti, and O. H. Turnbull (Eds.), *An Interdisciplinary Approach to the Rehabilitation of the Neurological Patient: A Cognitive Perspective* (pp. 33–46). Pavia: PI-ME Press.

Turnbull, O. H. (1997). Neglect: Mirror, mirror, on the wall—is the left side there at all? *Current Biology, 7*: 709–711.

Turnbull, O. H. (2000). Personal memories of experimental psychology and psychoanalysis. *Neuro-Psychoanalysis, 2*: 258–259.

Turnbull, O. H. (2001). Cognitive neuropsychology comes of age. *Cortex, 37*: 445–450.

Turnbull, O. H., and Owen, V. (in press). Affect in denial of deficit (anosognosia): A neurological patient overcome with one class of emotion. *Neuro-Psychoanalysis.*

Walsh, K. W. (1985). *Neuropsychology: A Clinical Approach.* Edinburgh: Churchill-Livingstone.

Weiskrantz, L. (1986). *Blindsight.* Oxford: Oxford University Press.

Yu, C. K. (2000). Clearing the ground: Misunderstanding of Freudian dream theory. *Neuro-Psychoanalysis, 2*: 212–213.

Yu. C. K. (in press). Neuroanatomical correlates of dreaming: The supramarginal gyrus controversy (dream-work). *Neuro-Psychoanalysis, 3.*

Zeki, S. (1993). *A Vision of the Brain.* Oxford: Blackwell.

Zueler, M., and Maas, J. (1994). An integrated conception of the psychology and biology of supergo development. *Journal of the American Academy of Psychoanalysis, 22*: 195–209.

INDEX

ABOUT THE AUTHORS

MARK SOLMS is a neuropsychologist and a psychoanalyst. His main focus in neuropsychology has been the brain mechanisms of dreaming. His findings are summarized in a book *The Neuropsychology of Dreams* (1997) and a target article in *Behavioral & Brain Sciences* (2000). He is also involved in scholarly research into the neurological origins of Freud's psychoanalytic concepts, and is editor and translator of *The Complete Neuroscientific Works of Sigmund Freud.* (He is also general editor of the Revised Standard Edition of the *Complete Psychological Works of Sigmund Freud.*) Mark Solms links his two interests by developing research methods to make psychoanalytic theories and concepts accessible to neuroscientific research. His main work to date in that direction is described in a book written with his wife and colleague, Karen Kaplan Solms, *Clinical Studies in Neuro-Psychoanalysis* (2000). He is also coeditor of an interdisciplinary journal for psychoanalysis and the neurosciences, *Neuro-Psychoanalysis.*

OLIVER TURNBULL is a Cambridge-trained neuropsychologist. He has published widely in neuroscientific journals, primarily on the topics of visuo-spatial disorders, laterality, and neuropsychological disorders involving false beliefs such as anosognosia and confabulation. He is currently Senior Lecturer in the Centre for Cognitive Neuroscience, University of Wales, Bangor. He is also the Secretary of the International Neuro-Psychoanalysis Society, and Research Digest Editor of the interdisciplinary journal *Neuro-Psychoanalysis.*